Intermediate Science and Theory

Electrical Installation Series – Intermediate Course

J.N. Hooper
M. Doughton
C. Duncan
E.G. Patterson
E.G. Stocks

Edited by Chris Cox

DELMAR
CENGAGE Learning™

Australia • Brazil • Japan • Korea • Mexico • Singapore • Spain • United Kingdom • United States

Intermediate Science and Theory
J.N. Hooper, M. Doughton, C. Duncan,
E.G. Patterson and E.G. Stocks

Publishing Director: John Yates
Publisher: Melody Dawes
Manufacturing Manager: Helen Mason
Senior Production Controller: Maeve Healy
Marketing Manager: Jason Bennett

For product information and technology assistance,
contact **emea.info@cengage.com**.

For permission to use material from this text or product,
and for permission queries,
email **clsuk.permissions@cengage.com**

British Library Cataloguing-in-Publication Data
A catalogue record for this book is available from the British Library.

ISBN: 978-1-86152-665-6

Cengage Learning EMEA
High Holborn House, 50-51 Bedford Row
London WC1R 4LR

Cengage Learning products are represented in Canada by Nelson Education Ltd.

For your lifelong learning solutions, visit
www.cengage.co.uk

Printed by Seng Lee Press, Singapore
4 5 6 7 8 9 10 – 10 09 08

About this book

"Intermediate Science and Theory" is one of a series of books published by Cengage Learning related to Electrical Installation Work. The series may be used to form part of a recognised course or individual books can be used to update knowledge within particular subject areas. A complete list of titles in the series is given below.

Electrical Installation Series

Foundation Course

Starting Work
Procedures
Basic Science and Electronics

Supplementary title:
Practical Requirements and Exercises

Intermediate Course

The Importance of Quality
Stage 1 Design
Intermediate Science and Theory

Supplementary title:
Practical Tasks

Advanced Course

Advanced Science
Stage 2 Design
Electrical Machines
Lighting Systems
Supplying Installations

Acknowledgements

The authors and publishers gratefully acknowledge the following illustration sources:

British Standards Institution for Figure 8.11; Brook Hansen plc for Figures 6.5, 6.34, 6.62, 6.63 and 6.64; Robin Instruments for Figure 10.12.

Every effort has been made to trace all copyright holders but if any have been inadvertently overlooked, the publishers will be pleased to make the necessary arrangements at the first opportunity.

Study guide

This studybook has been written to enable you to study either in a classroom or in an open or distance learning situation. To ensure that you gain the maximum benefit from the material you will find prompts all the way through that are designed to keep you involved with the subject. The book has been divided into parts each of which may be suitable as one lesson in the classroom situation. Certain parts of this book may be combined in one lesson period but this will depend upon the duration of the lesson. However if you are studying by yourself the following points may help you.

☞ Work out when, and for how long, you can study each week. Complete the table below and from this produce a programme so that you will know approximately when you should complete each chapter and take the progress and end tests. Your tutor may be able to help you with this. It may be necessary to reassess this timetable from time to time according to your situation.

☞ Try not to take on too much studying at a time. Limit yourself to between 1 hour and 2 hours and finish with a task or the self assessment questions (SAQ). When you resume your study go over this same piece of work before you start a new topic.

☞ You will find the answers to the questions at the back of the book but before you look at the answers check that you have read and understood the question and written the answer you intended.

☞ A "progress check", at the end of Chapter 5, and "end tests" covering all the material in this book are included so that you can assess your progress.

☞ Tasks are included where you are given the opportunity to ask colleagues at work or your tutor at college questions about practical aspects of the subject. These are all important and will aid your understanding of the subject.

☞ As you progress through this book you will come across new terms and new formulae. Make a note of these at the end of this book in the space provided and include the page number and a definition and/or comment as appropriate. You can then use this as a handy reference guide.

☞ It will be helpful to have available for reference a current copy of BS 7671 when studying this book.

☞ Your safety is of paramount importance. You are expected to adhere at all times to current regulations, recommendations and guidelines for health and safety.

Study times						
	a.m. from	to	p.m. from	to		Total
Monday						
Tuesday						
Wednesday						
Thursday						
Friday						
Saturday						
Sunday						

Programme	Date to be achieved by
Chapter 1	
Chapter 2	
Chapter 3	
Chapter 4	
Chapter 5	
Progress check	
Chapter 6	
Chapter 7	
Chapter 8	
Chapter 9	
Chapter 10	
End test	

Contents

8 Illumination   143

9 Efficiency, Work, Energy and Power   165

10 Instruments   175

Short answer end test   189

Multi-choice end test   193

Answers   195

1

Magnetism

Complete the following to remind yourself of some important facts on this subject that you should remember from previous studies on this subject.

Name two basic types of permanent magnet.

Magnetic fields are invisible. How are they normally represented on a diagram of a magnetic circuit?

Name three types of electromagnetic device.

What is induced into a conductor when it cuts through the lines of flux of a magnetic field?

On completion of this chapter you should be able to:

- describe, with simple sketches, the pattern and direction of magnetic flux paths
- state and apply the units, symbols and quantities for magnetic flux, magnetic flux density and magnetomotive force
- determine the force on a current carrying conductor in a magnetic field
- determine the e.m.f. induced in a conductor moving through a magnetic field
- explain how an e.m.f. may be produced by self induction and mutual induction
- describe the effects of (a) hysteresis loss (b) eddy current loss and methods of reducing these losses
- describe the basic principle of operation of d.c. machines, transformers and relays
- state the principle of operation of overload protective devices

Part 1

Magnetic fields

A magnet will affect the space around it in such a way that any other magnets placed in this space experience forces. The actual space in which these forces occur is known as the magnetic field. The presence of a magnetic field surrounding a bar magnet can be demonstrated by sprinkling iron filings on to a sheet of paper (or thin card) placed on top of the magnet, and when it is gently tapped the iron filings take up the field pattern as shown in Figure 1.1.

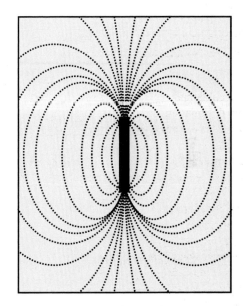

Figure 1.1   Magnetic field pattern

It can clearly be seen that the iron filings produce a very random field pattern and that the flux does not exist as a number of separate lines. However, it is easier to explain and illustrate the various magnetic effects by representing magnetic field patterns as a number of separate lines of flux. This basic concept will be used throughout the text.

Consider the magnetic field pattern and flux paths of a permanent "bar" magnet and an electromagnet (solenoid) as shown in Figures 1.2 and 1.3.

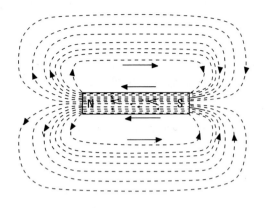

Imaginary lines of magnetic flux

Figure 1.2   *Bar magnet field pattern*

The flux paths that the lines of flux take are through the magnet and the space around the magnet. The lines of flux actually "leave" the north pole and enter the south pole of the magnet.

The concentration (density) of the magnetic flux is strongest within the magnetic material since there are more lines of flux *closer together* than there are in the space surrounding the magnet.

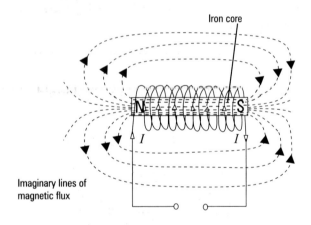

Iron core

Imaginary lines of magnetic flux

Figure 1.3   *Solenoid field pattern*

It can clearly be seen that the field patterns and flux paths are very similar. The only difference being that the magnetic field of the solenoid is produced by the current flowing through the coil.

Looking at the diagrams more closely it can be seen that the imaginary lines of magnetic flux have very distinct properties:
- they always form complete closed loops
- they never cross one another
- they have a definite direction (North to South)

Two other properties which are not so distinct on the diagrams are:
- they try to contract as if they were stretched elastic threads
- they repel one another when lying side by side, and having the same direction

Now let's consider the magnetic field patterns due to two bar magnets placed side by side.

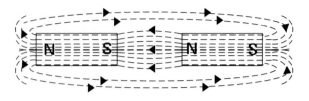

Figure 1.4   *Attraction – note: loops have been cut to save space*

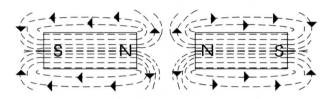

Figure 1.5   *Repulsion*

Figure 1.4 clearly shows that when unlike poles are adjacent, the magnets are attracted to each other and Figure 1.5 shows that when like poles are adjacent the magnets are repelled from each other.

Concentric magnetic field directions

When an electric current flows through a conductor it produces a magnetic field round the conductor in the form of concentric rings as shown in Figures 1.6 and 1.7.

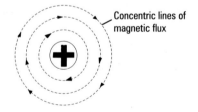

Concentric lines of magnetic flux

Figure 1.6   *Current flowing away from you (cross) – concentric field direction is clockwise*

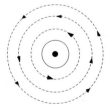

Figure 1.7   *Current flowing towards you (dot) – concentric field direction is anticlockwise*

The current flow direction can easily be remembered by the Right-Hand Grasp Rule.

Figure 1.8   *The fingers point in the direction of the
                    magnetic field and the thumb in the direction of
                    the current flowing*

Alternatively apply Maxwells Corkscrew Rule, which states that, if a normal right hand thread screw is driven along the conductor in the direction taken by the current, its direction of rotation will be the direction of the magnetic field.

Magnetic flux paths associated with adjacent parallel current carrying conductors

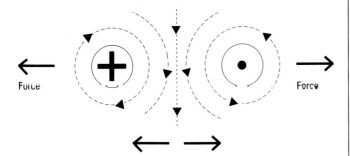

Figure 1.9   *Here the lines of flux tend to "repel" each other
                    as they are in the same direction. The
                    conductors are forced away from each other.
                    (REPULSION)*

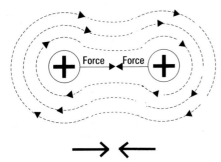

Figure 1.10   *Here the lines of flux "cancel" each other out.
                    The conductors are now pulled towards each
                    other. (ATTRACTION)*

Remember that a "cross", ✛, indicates the current is flowing away from you and a "dot", ●, it is flowing towards you.

Try this

Draw on the diagram below:

1. The resultant magnetic flux paths around the conductors and their directions.

2. The direction of the forces acting on each conductor.

Figure 1.11

Remember

The Right Hand Grasp Rule is used to find the relative direction of the current flowing and the magnetic field set up by the current.

Right Hand Grasp Rule applied to a solenoid

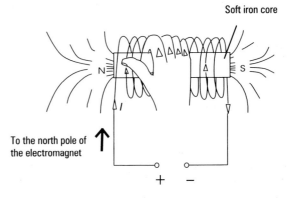

Figure 1.12   *Right hand grasp rule*

The fingers point in the direction of the current flowing and the thumb indicates the direction of the magnetic field.

Note: The thumb points towards the North Pole of the electromagnet.

Also if the current direction is reversed then the field direction is also reversed.

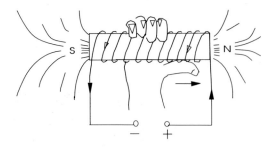

Figure 1.13

Magnetic fields/flux paths

The magnetic field of a permanent magnet or an electromagnet is made up of many lines of _____.

Lines of magnetic flux never _____ and have a definite _____.

Unlike poles _____ and like poles _____ each other.

An electric current flowing through a conductor produces a _____ round the _____ in the form of _____.

The fingers point in the direction _____, and the thumb points in the direction _____ when applying the R.H. grasp rule to a straight conductor.

When two adjacent parallel _____ conductors have currents flowing in opposite directions the lines of _____ tend to_____ each other.

When applying the R.H. grasp rule to a solenoid the _____ point in the direction of the current flowing and the _____ indicates the direction of the _____.

Try this

Redraw the diagrams shown below, and show the resultant magnetic field pattern in each case.

a.  b. 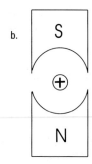 c.

Figure 1.14

Try this

Draw on the diagram below:
(a) The concentric lines of magnetic flux around each of the top and bottom coil loops.
(b) The pattern and direction of the resultant magnetic flux paths around the top and bottom sides of the coil. Indicate the north and south poles of the solenoid.

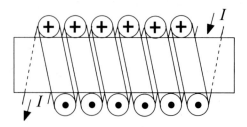

Figure 1.15

Part 2

Magnetic field properties

Magnetic flux
(quantity symbol Φ, unit symbol Wb)

Magnetic flux is a measure of the magnetic field produced by a permanent magnet or electromagnet. The unit of magnetic flux is the Weber (Wb) which is pronounced "vayber".

Magnetic flux density
(quantity symbol B, unit symbol T)

Magnetic flux density depends on the amount of magnetic flux (i.e. the number of lines of flux) which is concentrated in a given cross-sectional area of the flux path. The strength of a magnetic field is measured in terms of its flux density. The unit of magnetic flux density is the tesla (T) and

$$1 \text{ tesla} = 1 \text{ weber per square metre.}$$

This may be found by using the formula:

$$B = \frac{\Phi}{A} \text{ where } A \text{ is the area}$$

$$\text{Therefore 1 tesla} = \frac{1 \text{ weber}}{1 \text{ square metre}}$$

$$\text{or 1 T} = 1 \text{ Wb/m}^2$$

Figure 1.16 clearly shows that magnetic flux density is a measure of the amount of flux (Φ) which exists within an area (A) perpendicular to the direction of the flux.

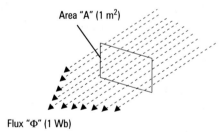

Flux "Φ" (1 Wb)

Figure 1.16   Cross-section through a uniform magnetic flux

Try this
Calculate the flux density existing in an area of 10 m² if a uniform magnetic flux of 1 Wb exists at right angles to that area.

Now it can be seen (Figure 1.17) for the same amount of flux spread throughout a larger area that the flux density is reduced (in this case by one tenth of what it was before).

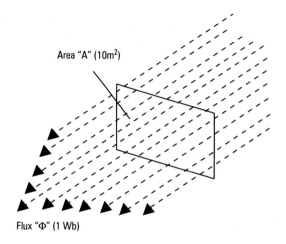

Area "A" (10m²)

Flux "Φ" (1 Wb)

Figure 1.17   Area 10 × greater, Flux density 10 × less

Magnetomotive force, abbreviation m.m.f.
(quantity symbol F, unit symbol A)

Magnetomotive force is the "force" which produces a magnetic flux and its magnitude depends on the number of turns on the coil (N) and on the current in the coil (I).

Hence:

magnetomotive force = number of turns × current

or $F = NI$

The unit of m.m.f. is the ampere-turn (At); however, the number of turns has no units so the unit for m.m.f. is simply the ampere (A).

Example

Calculate the m.m.f. produced by a current of 2 A flowing in a 400-turn coil.

$$\begin{aligned} F &= NI \\ &= 400 \times 2 \\ &= 800 \text{ At} \end{aligned}$$

Try this
1. A solenoid is wound with 2000 turns of wire and a current of 500 mA is passed through the wire. Find the magnetomotive force.

2. A magnetomotive force of 1200 At is required from a coil having 2400 turns of wire. How much current must be passed through the wire?

Magnetic field strength

The strength of a magnetic field produced by a solenoid (for example) basically depends on

- the magnitude of the current flowing through the solenoid's coil
- the number of turns on the coil
- the type of core material used

The magnetic field strength can be increased by

- increasing the current flowing through the coil
- winding more turns on the coil
- selecting a suitable core material (with improved magnetic properties)

Try this

Draw on the diagrams below the resultant magnetic flux paths and indicate the electromagnet with the greatest magnetic field strength with more lines of flux. State why this electromagnet has a greater field strength.

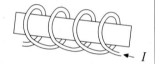

Figure 1.18   *Figure 1.19*

Points to remember   ◄--------------

Magnetic field properties

Magnetic flux (quantity symbol _____) is a measure of the _____ and its unit is the _____

Magnetic flux density (quantity symbol ____) depends on the amount of _____ and its unit is the _____

Magnetomotive force, abbreviation _____ is the _____ which produces a _____, and its magnitude depends on the _____(N) and on the _____(*I*).

The unit of magnetomotive force is the _____

Magnetic field strength can be increased by _____ the current flowing through the coil, or by winding _____ on the coil.

Part 3

Electromagnetic force on a current-carrying conductor in a magnetic field

Nearly all motors work on the basic principle that when a current-carrying conductor is placed in a magnetic field it experiences a force. This electromagnetic force is shown in Figures 1.20, 1.21 and 1.22.

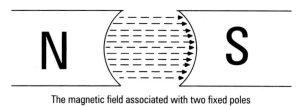

The magnetic field associated with two fixed poles

The magnetic field associated with a current carrying conductor

The current is going away from you, therefore the field is clockwise (corkscrew rule).

Figure 1.20

Remember
A "cross" **+** indicates the current is flowing away from you and a ● that it is flowing towards you.

Now let's place the current-carrying conductor into the magnetic field. (Figure 1.21)

Direction of the resultant force on the conductor

The resultant magnetic field when the current-carrying conductor is placed between the fixed poles.

Figure 1.21

- the main field now becomes distorted
- the field is weaker below the conductor due to the fact that the two fields are in opposition
- the field is stronger above the conductor because the two fields are in the same direction and aid each other. Consequently the force moves the conductor downwards.

If either the current through the conductor or the direction of the magnetic field between the poles is reversed, the force acting on the conductor tends to move it in the reverse direction. Figure 1.22

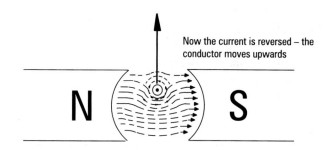

Now the current is reversed – the conductor moves upwards

Figure 1.22

The direction in which a current-carrying conductor tends to move when it is placed in a magnetic field can be determined by Fleming's left hand rule.

Fleming's Left Hand (motor) rule

If the first finger, the second finger and the thumb of the left hand are held at right angles to each other, Figure 1.23, then with the **f**irst finger pointing in the direction of the **f**ield (N to S), and se**C**ond finger pointing in the direction of the **C**urrent in the conductor, then the thu**m**b will indicate the direction in which the conductor tends to **m**ove.

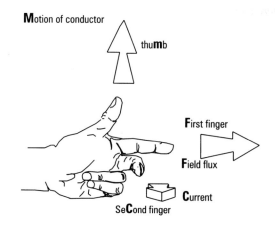

Motion of conductor

thu**m**b

First finger

Field flux

Current

Se**C**ond finger

Figure 1.23

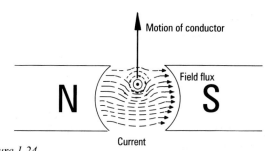

Motion of conductor

Field flux

Current

Figure 1.24

7

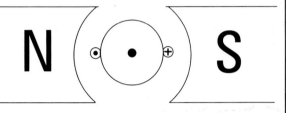
Calculating the force on a current-carrying conductor in a magnetic field

The magnitude of the force on a current-carrying conductor can be calculated by the formula:

$$F = BIl$$

Where
  F = force in newtons (N)
  B = flux density in tesla (T)
  l = length of conductor affected by the magnetic field in metres (m)
and
  I = current in amperes (A)

Note that the formula $F = BIl$ only applies to conductors which are moving at **right angles** to the magnetic field.

Example

A conductor 0.3 m long carries a current of 30 A at right angles to a magnetic field of flux density 1.5 T. Determine the force exerted on the conductor.

$$
\begin{aligned}
F \quad &= BIl \\
&= 1.5 \times 0.3 \times 30 \\
&= 13.5 \text{ N}
\end{aligned}
$$

Points to remember ◄-------------------

Force on a current-carrying conductor

Nearly all motors work on the basic principle that when a _____ is placed in a _____ it experiences a _____

A "cross", +, indicates _____

and a "dot", ●, indicates _____

Fleming's Left Hand Rule is used to determine the _____ when placed in a _____

When applying Fleming's Left Hand Rule the first finger indicates _____, the second finger _____ and the thumb _____.

To calculate the force on a current carrying conductor when at _____ to a magnetic field apply the formula _____.

Part 4

Induced e.m.f.

A straight conductor passing between the poles of a magnet has an e.m.f. induced in it which is equivalent to the product of the flux density, the length of conductor in the field and its velocity.

$$E = B\,l\,v$$

Where E = induced e.m.f. in volts   V
         B = flux density in tesla   T
         l = length in metres   m
         v = velocity in metres/sec   m/s

The direction of the induced e.m.f. can be determined by Fleming's right hand rule.

Fleming's right hand rule

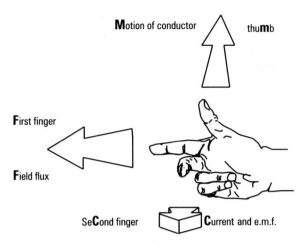

Figure 1.26   Fleming's Right Hand Rule

Note: The formula $E = Blv$ can only be applied when the conductor is moving at right angles to the magnetic field, as shown in Figure 1.27.

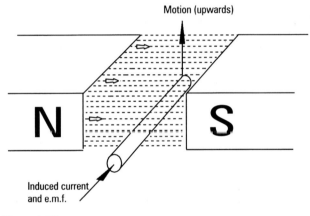

Figure 1.27

Example
When a conductor 0.2 m long is moved at a velocity of 12 m/s through a magnetic flux which has a strength of 0.9 T it produces an e.m.f. of

$E = Blv$
$E = 0.2 \times 12 \times 0.9$
$E = 2.16\,\text{V}$

Try this
Calculate the e.m.f. induced in a conductor of length 30 cm which is moving through a magnetic field of flux density 0.5 T at a velocity of 40 m/s.

Induced e.m.f. by rate of change of magnetic flux

$$E = \frac{N(\Phi_2 - \Phi_1)}{t}$$

Where
E = induced e.m.f. in volts V
N = number of coil turns
Φ_1 = initial flux in webers Wb
Φ_2 = final flux in webers Wb
t = time taken for change in flux (from Φ_1 to Φ_2) s

Example
The flux linking with a 600-turn coil changes from 60 mWb to 120 mWb in 100 ms. Calculate the induced e.m.f. in the coil.

$$E = \frac{N(\Phi_2 - \Phi_1)}{t}$$

$$= \frac{600\,(120 - 60)}{100}$$

$$= \frac{600 \times 60}{100}$$

$$= 360\,\text{V}$$

Try this
A flux linking with a 300-turn coil changes from 2 mWb to 4 mWb in 0.02 s. Calculate the magnitude of the induced e.m.f.

Self inductance

As discussed earlier, a current-carrying conductor possesses its own magnetic field. When this conductor is wound into a coil, the field produced has the properties of a magnet. This becomes more pronounced if the coil is wound around an iron core. When the current is first switched on the magnetic field expands outwards, cutting the conductors and producing an induced e.m.f. This has the effect of opposing the e.m.f. producing the rise in flux density.

A German physicist, Heinrich Lenz said the same thing (in German) in about the year 1834 and this became known as Lenz's Law:

"The direction of an induced e.m.f. is always such that it tends to set up a current opposing the motion or the change in flux responsible for producing that e.m.f."

An inductor is said to have an inductance of **one henry** when a current which is changing at the rate of **one ampere per second** produces an induced e.m.f. of **one volt**.

The henry is the unit of inductance and you will find it used to evaluate the inductive properties of chokes and coils in a variety of applications. One henry is quite a large value in practical terms and it is usual to find inductors measured in millihenries for everyday applications.

When an inductive circuit is first switched on, the induced e.m.f opposes the increase in current. Instead of immediately rising to its final value, the current rises more gradually, at a rate governed by the inductance of the circuit.

For example, a circuit having an inductance of 250 mH and a resistance of 250 mΩ is connected across a 1V supply.

The final circuit current is

$$I \quad = \frac{U}{R}$$
$$= \frac{1}{0.25}$$
$$= 4 \text{ A}$$

The only difference between this and a purely resistive circuit is the time taken for the current to reach this value.

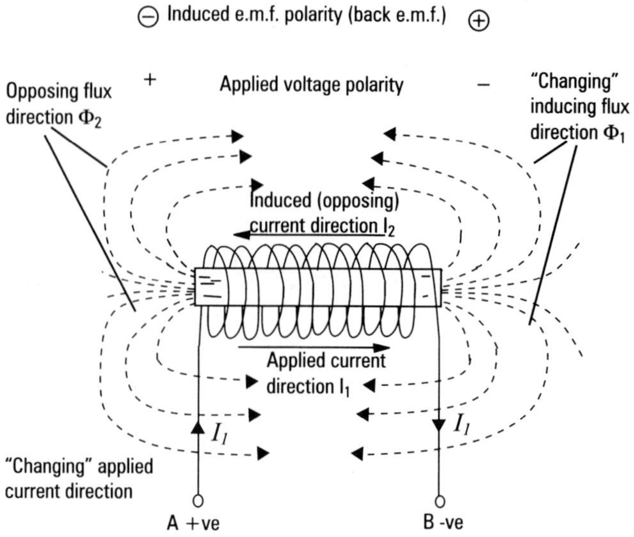

Figure 1.28   *Lenz's Law*

In Figure 1.28 the inducing flux (Φ_1) is steadily increased by increasing the applied voltage to terminals A and B, and the applied current (I_1). Due to this increasing flux Φ_1 an e.m.f. is induced in the coil of opposite polarity to the applied voltage. This e.m.f. will cause the induced current (I_2) to flow through the coil in an opposite direction to the applied current (I_1). This induced current (I_2) will set up a flux (Φ_2) in such a direction as to oppose the original inducing flux (Φ_1).

Mutual inductance

The transformer is a static electromagnetic device which operates on the principle of mutual inductance. Having studied electromagnetism you will be aware that a coil which is situated in its own magnetic field is capable of producing an induced e.m.f. due to its inductive properties.

If two coils share the same magnetic field, an e.m.f. will be induced in both coils. When a change in current in the first coil induces an e.m.f. in the second they are said to be mutually inductive.

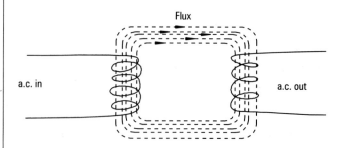

Figure 1.29

Try this

Explain briefly how an e.m.f. will be produced in both coils (A and B) on the diagram shown below.

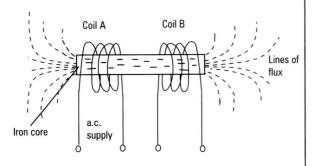

Coil A   Coil B

Lines of flux

Iron core

a.c. supply

Figure 1.30

Points to remember ◀ – – – – – – – – – – – –

Induced e.m.f./self and mutual induction

To calculate the induced _____ in a conductor moving at _____ to a magnetic field, apply the formula _____

When applying _____ Right Hand Rule the first finger indicates the _____, the second finger the _____ and the thumb the direction that _____.

The magnitude of an _____ due to a rate of change of _____ can be found by the formula

$E =$ _____

Lenz's Law states that: The _____ of the induced e.m.f. in a circuit is such that it sets up a _____ which _____ the motion or change of _____ producing that _____ e.m.f.

An e.m.f. may be produced by _____ induction or _____ induction.

An inductor has an inductance of _____ when a current which is changing at the rate of_____ produces an induced e.m.f. of _____

Part 5

Energy stored in an inductive circuit

During the time taken for an inductive circuit to reach maximum current after switching on, energy from the electrical input has been transferred to the magnetic field. Whilst the current is flowing, this energy level is maintained and held in storage in the inductor. The amount of energy is determined by the current and the inductance and can be calculated by;

Energy $W = \dfrac{1}{2}LI^2$ Joules

Example

A coil of 250 mH carries a d.c. current of 4 A. How much energy is stored in the coil while this current is flowing?

$$
\begin{aligned}
\text{Energy } W \quad &= \frac{1}{2}LI^2 \text{ Joules} \\
&= 0.5 \times 250 \times 10^{-3} \times 4^2 \\
&= 0.125 \times 16 \\
&= 2 \text{ Joules}
\end{aligned}
$$

The presence of this stored energy does not normally cause a problem. Indeed it is the stored energy in an inductor that helps to smooth the output of many power supply units.

The problem arises in the switching of inductive circuits.

When switched off, the inductor releases its energy back into the circuit in the form of a high induced voltage. This is exactly what occurs when the contact breaker points open, a surge of high voltage is directed to the spark plugs and the resultant spark discharge ignites the fuel mixture.

Imagine what would happen if a voltage of many thousands of volts suddenly appeared amongst a load of delicate electronic components. That's when special precautions have to be taken to suppress this induced voltage before it can do any damage.

Remember

Energy is being stored in the magnetic field when the field is being "built up" and the energy is released into the electrical circuit when the field "collapses".

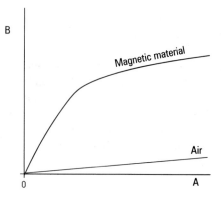

Figure 1.32   *Typical magnetisation (B/H) characteristics for air and a magnetic material*

The magnetisation curve

This is also known as the B/H curve and shows the relationship between flux density and magnetising force for a specific type of magnetic material.

The quantity H, i.e. the magnetising force, is the ampere turns per metre length of the magnetic path.

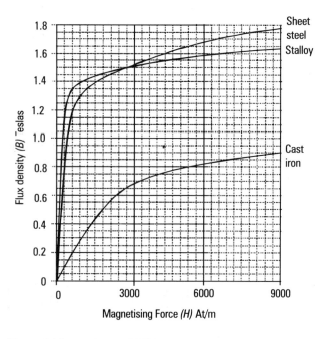

Figure 1.31   *A typical B/H curve*

You will see that the tops of the curves flatten out. This is because the core becomes "saturated" and cannot carry any more magnetic flux even though the current in the coil carries on increasing. Some materials saturate before others, for example cast iron, and some are more easily magnetised (more permeable) for example stalloy. These features can be seen on the B/H curve showing comparative examples, Figure 1.31.

Note: If the magnetic circuit consists entirely of air, or any other non-magnetic material then, the resulting graph will be a straight line as shown in Figure 1.32.

Hysteresis

As the current is increased the flux density increases. As the current is decreased, the flux density decreases but it does not follow the same path. On the downward path the flux density is higher than on the way up.

When an electromagnet is connected to an a.c. source the flux in the core is increased then decreased and reversed and brought back to zero again in the course of each cycle of the supply. This gives the characteristic form known as the hysteresis loop as shown here.

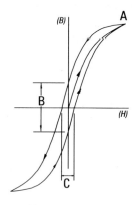

Figure 1.33   *Typical hysteresis loop*

Note: For a 50 Hz supply, 50 loops will be produced each second.

There are several features of the loop which can convey information about the properties of the magnetic material.

A. SATURATION
    The flux density increases as the magnetising force is increased but at some stage, the core begins to saturate and when fully saturated, the flux density will not increase any further, no matter how much current flows in the coil.

B. REMANENCE
    When the magnetising current reduces to zero, there will still be some flux remaining in the core. This will vary with the type of magnetic core material. Permanent magnets will have a high "remanent" (remaining) flux

but in softer magnetic materials the amount of flux remaining will be much smaller.

C. COERCIVE FORCE

Before the remanent flux can be removed, the magnetising current has to be reversed. This can be seen on the hysteresis loop as the negative magnetising force required before the flux density passes through zero. The coercive force varies enormously between permanent magnetic materials and those, for example, used in transformer cores.

Losses in magnetic materials

When a magnetic material is subjected to alternations of magnetic flux (flux reversals) some energy is lost as work is done in magnetising and demagnetising the material. This loss is referred to as hysteresis loss, and it is proportional (per cycle of the a.c. supply) to the area of the hysteresis loop. (Figure 1.33). It can clearly be seen from the diagram that the "fatter" the loop is the greater the energy loss due to hysteresis and with a "thinner" loop that the loss is less.

This loss occurs in transformers due to the flux increasing and decreasing in the transformer core. It also occurs in the core of a generator or motor due to the rotor passing first under a north pole, then under a south pole, and thus experiencing alternations of magnetic flux.

Another form of energy loss which occurs in transformer, generator and motor cores is caused by the e.m.f. which is induced in the core due to the changing magnetic flux (produced by the alternating current in the windings) cutting through the core.

This induced e.m.f. causes "circulating" currents to flow in the core, called "eddy currents". These eddy currents cause unnecessary heating and power losses in the core, i.e. the larger the current the more the heat dissipation. The energy loss associated with these currents is known as eddy-current loss and may be reduced by laminating the cores. (This will be covered in more detail in Chapter 6.)

Magnetic materials

Magnetic materials can broadly be divided into two classes:
- "soft" magnetic materials, which gain and lose their magnetism easily
- "hard" magnetic materials, which are comparatively difficult to magnetise but strongly retain their magnetism.

Applications of magnetic materials

"Soft" magnetic materials are used, for example, to construct transformer cores, stator and rotor cores of electric motors. Examples of such materials are silicon steel and stalloy (a silicon-iron alloy), with a hysteresis loop similar to that shown in Figure 1.34b.

"Hard" magnetic materials are used in the construction of permanent magnets. Examples of such materials are hard steel and ALNICO (an alloy of aluminium, nickel, iron, copper and cobalt), with a hysteresis loop similar to that shown in Figure 1.34a.

Ferrite cores are used for audio and high frequency applications and have a hysteresis loop similar to that shown in Figure 1.34c.

Ferrite (dust) cores are made from ceramic-like magnetic substances made from oxides of iron, nickel, cobalt, magnesium, aluminium and manganese.

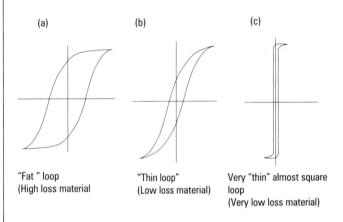

(a) (b) (c)

"Fat" loop (High loss material)   "Thin loop" (Low loss material)   Very "thin" almost square loop (Very low loss material)

Figure 1.34   Hysteresis loops for different materials.

Note: Hysteresis loss may be kept to a minimum by choosing the most suitable ferromagnetic material for the particular application.

Try this

Sketch a typical hysteresis loop for a ferro-magnetic material used for constructing a transformer core.

Part 6

Basic principle of operation of a d.c. motor

Magnetic materials, energy, hysteresis and eddy current losses

Energy is stored in the _____ _____ of an inductor when the field _____ _____ , and is released into the electrical circuit when the field _____.

The B/H characteristic for a _____ _____ is the same as that for air.

Soft magnetic materials _____ and _____ their magnetism easily.

Hard magnetic materials _____ their magnetism.

A permanent magnet material will have a _____ hysteresis _____ , and a stator core of an electric motor will have a _____ loop because it is made from a_____ magnetic material.

Circulating currents are called _____ currents which are induced into the _____ of a transformer and may be reduced by _____ __ _____

Ferrite cores are used for _____ and _____ _____ applications and have a _____ _____ hysteresis loop which is almost _____.

When a magnetic material is subjected to flux reversals some _____ is lost as _____ is done in magnetising and _____ the material, this loss is called _____ _____.

Rotation

Consider the single-turn coil (which is free to rotate on a shaft) placed inside the fixed magnetic field. Figure 1.35

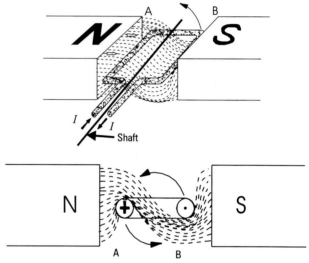

Figure 1.35   *The rotating force produced on a single coil in a magnetic field*

Applying Flemings L.H. Rule, coil side "A" will have a downwards force produced on it, and coil side "B" will have an upwards force produced on it. The result of the forces on coil sides A and B, is to produce a torque in an anticlockwise direction and thus produce rotary motion in this direction. **Note:** the coil will only be forced to a position at 90° to the fixed magnetic field as shown in Figure 1.36.

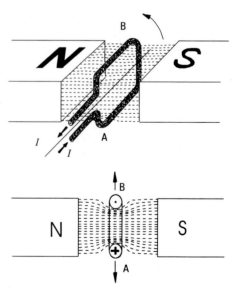

Figure 1.36   *The coil has now rotated 90 °*

Commutation

To get past this point a second coil is used placed at 90° to the first. Current is switched off to the first coil and on to the second. The force on the second coil now rotates the assembly until it is at 90° to the fixed magnetic field. If the direction of rotation is to be maintained the current has to be switched off in the second coil and back on in the first but now in the reverse direction. This switching is carried out automatically with a rotating switch called a commutator. Figure 1.37

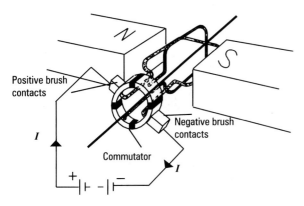

Figure 1.37   *A simple armature arrangement with a four segment commutator*

The practical d.c. motor has many armature coils (windings) and a multisegment commutator Figure 1.38.

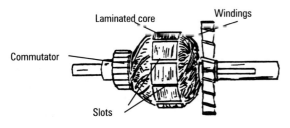

Figure 1.38   *Armature*

Note: The commutator segments are insulated from the shaft and each other.

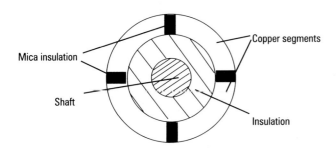

Figure 1.39   *Multisegment commutator*

Try this

Assuming that the coil, Figure 1.35, has rotated 180° from its original position indicate on the diagram below

a)   The coil sides A and B
b)   the direction of the current induced in each coil side when supplied from the battery in Figure 1.37.
c)   The main field between the two poles
d)   The direction of the forces acting on each coil side
e)   The direction of rotation
f)   The position that the coil will move to, due to the forces in (d) above.

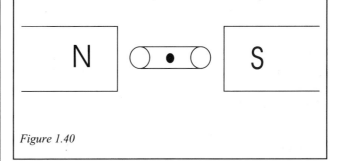

Figure 1.40

Basic principle of operation of a d.c. generator

It is important to note that the direct-current generator is basically the same as the direct-current motor. Both machines are energy convertors. The motor converts electrical energy into mechanical energy, and the generator converts mechanical energy into electrical energy.

When the single-turn coil is rotated within the magnetic field, Figure 1.41, an alternating "voltage" is produced. This alternating voltage is converted to a direct one by the commutator. The commutator rotates with the coil so that the two segments continually interchange with the two carbon brushes, which are stationary. Each end of the coil is connected to a segment of the commutator.

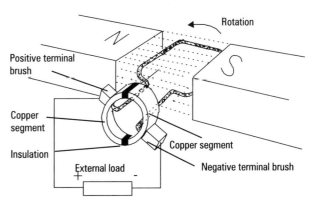

Figure 1.41   *Simple (single-loop) generator*

Generator output

The generated e.m.f. in the coil alternates but the output voltage at the brushes retains the same polarity. This is because after 180° rotation the segments interchange their brushes so that as the voltage begins to increase again in the opposite direction the polarity of the output voltage remains the same (Figure 1.42).

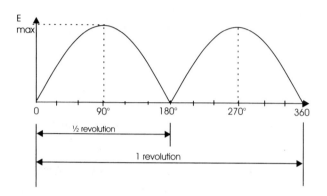

Figure 1.42   *Generator output voltage waveform*

Remember
The purpose of the commutator:
Motor

to transfer the supply current to the armature coils via brushes and to reverse the direction of the current flowing in the coils as the armature rotates.

Generator

to convert the alternating voltage and current induced in the armature coils into a direct voltage and current at the brushes which are connected to the external circuit.

Let's now consider the output from the generator in a little more detail by applying Flemings Right-Hand (generator) Rule.

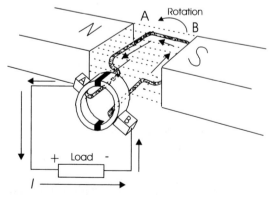

Figure 1.43

Applying Flemings R.H. Rule to coil in Figure 1.43 side "A" is moving downwards so the direction of its induced current is "outwards" and coil side "B" is moving upwards so the direction of its induced current is "inwards". Therefore the current flows out of the generator via brush "A", then through the load and back into the generator via brush "B".

Try this
Assuming that the coil, Figure 1.43, has rotated 180° from its original position

a)   indicate on the diagram
    i)   coil sides A and B
    ii) brushes A and B
    iii) the direction of the current induced in each coil side and the direction of the current in the external circuit
    iv) the polarity of the supply across the load

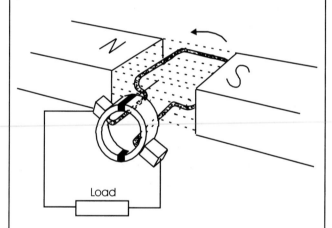

Figure 1.44

b)   complete the explanation given below:

Applying _____ , coil side _____ is moving _____ , so the direction of its induced current is _____ and, coil side _____ is moving _____ , so the direction of its induced current is _____ .

Therefore the current _____ out of the generator via _____ then through the _____ , and _____ into the generator via _____ .

Magnitude of generated e.m.f.

The magnitude of the generated e.m.f. depends on the position of the coil within the magnetic field. Figure 1.45

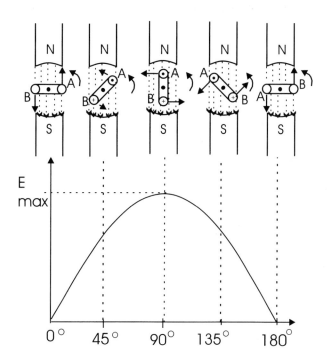

Figure 1.45   Magnitude of generated e.m.f. (During 180° rotation of the coil, and at intervals of 15°)

It can be seen from the diagram above that at:

0°   no magnetic flux is being cut, since the coil sides are moving **parallel** to the lines of magnetic flux, therefore no e.m.f. is generated

45°   some flux is being cut since the coil sides are now moving **at an angle** to the lines of flux therefore some e.m.f. is being generated

       Note: the magnitude of the e.m.f. depends upon the angle which the coil sides are moving through the lines of flux

90°   maximum flux is being cut since the coil sides are moving at 90° (**right angles**) to the lines of flux therefore maximum e.m.f. is generated

135°   magnitude of e.m.f. generated same as at 45°

180°   no e.m.f. generated

D.C. machines

A motor converts _____ into _____ and a generator converts _____ into

_____ .

The simple (single-turn coil) d.c. motor has _____ commutator segments and will only rotate to a position _____ to the fixed _____ .

To get past this point a _____ is used placed at_____ to the first _____ . This simple motor will have_____ commutator segments insulated from _____ and the _____ .

The commutator is like a _____ switch which _____ the direction of the _____ in the _____ coils of the motor.

The commutator on a d.c. generator converts the _____ induced in the _____ coils to a _____ .

Part 7

Basic principles of operation of transformers

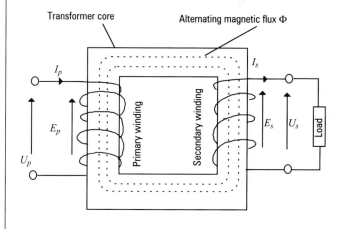

Figure 1.46   Double wound transformer

When an alternating voltage (U_P) is applied to the primary winding it causes an alternating current (I_P) to flow in the primary circuit. This current produces an alternating magnetic flux (Φ) in the core of the transformer. Due to this changing magnetic flux an alternating e.m.f. (E_P) is induced in the primary winding by means of "self induction" and alternating e.m.f. (E_S) is induced in the secondary winding by means of "mutual induction". It is this mutually induced e.m.f. (E_S) that is providing the output voltage (U_S) to the load and when the load is connected across the secondary winding the secondary circuit of the transformer is completed and alternating currents (I_P) and (I_S) will flow in the primary and secondary circuits.

Remember

The alternating magnetic flux links with both primary and secondary windings inducing voltages in each.

The induced e.m.f. (E_P) in the primary winding opposes the applied voltage (U_P) according to Lenz's Law.

Basic principle of operation of a relay

A relay is an electromagnetic device that opens and/or closes contacts to control the flow of current in one or more separate circuits. Figure 1.47

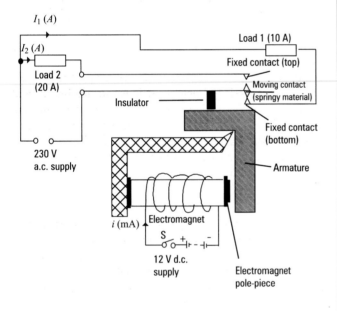

Figure 1.47   Simple electromagnetic relay

Operation

When switch "S" is closed, current "i" flows in the relay coil and the magnetic field set up by this current attracts the armature to the electromagnets pole-piece. The armature forces the moving contact "away" from the "bottom" fixed contact and "up" to the "top" fixed contact. In doing so current "I_1" ceases to flow through Load 1 and current "I_2" flows through Load 2. (It can clearly be seen that the relay's function in this circuit is as a "changeover" switch.)

Reasons for using a relay

There are many reasons for using a relay, some are given below (and are also included in the circuit diagram):

- Control circuits, carrying only a few milliamps, may be used to switch heavy loads carrying large currents.
- The control circuit may have a different type of supply to the main circuit.
- A low voltage control circuit can be used to switch a higher voltage main circuit.
- For safety reasons the control circuit may be "electrically isolated" from the main circuit.

Note: a practical relay will have many normally open (N/O) and normally closed (N/C) contacts.

Points to remember ◀‑ ‑ ‑ ‑ ‑ ‑ ‑ ‑ ‑ ‑ ‑ ‑

Transformers

The_____ (U_P) causes an _____ (I_P) to flow in the _____ circuit. This sets up an _____(Φ) in the transformer's _____. This flux_____ with both_____ and _____windings and _____ voltages into them.

The e.m.f. (E_P) is induced in the_____ by means of _____, and the e.m.f. (E_S) is induced in the _____ by means of _____.

Relays

The relay is an _____ device that _____ and/or _____to control the _____ in one or more separate _____. When current flows through the relay's_____ it sets up a _____ which attracts the _____ to the electromagnet's _____ and when the armature moves, it is designed to _____ and/or _____ contacts.

Principle of operation of overload (overcurrent) protective devices

Overload or overcurrent protection

All motors over 0.37 kW, either single or three-phase, must have control equipment with overcurrent protection. This is provided by either magnetic or thermal overload trips.

Magnetic overload trip

The simple type of magnetic overload trip consists of a solenoid coil inserted in the line circuit, which trips the starter mechanically, or through a contact which de-energises the control circuit. Figure 1.48

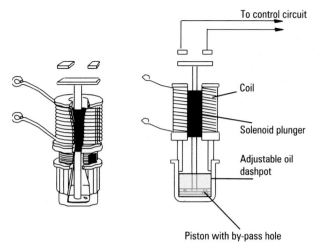

Figure 1.48

Labels in figure: To control circuit, Coil, Solenoid plunger, Adjustable oil dashpot, Piston with by-pass hole

The current setting can be adjusted by raising or lowering the dashpot. The time delay can be increased by reducing the piston by-pass hole diameter, and decreased by enlarging it.

To prevent this type of overload tripping immediately, for instance on starting when the motor is drawing excessive current, it is fitted with a time delay. Oil dashpot time lags, as shown in Figure 1.48, are frequently used but pneumatically operated types are also available.

Principle of operation

Excess motor current causes the solenoid plunger to lift but the movement is damped by the dashpot.

It takes time for the oil to leak through the piston by-pass hole (of metered diameter) and only when this process has run to completion can the trip contact close, to trip the control circuit.

The time-lag is usually fixed by the manufacturer and only the recommended fluid (dashpot oil) should be used to top up the dashpots. Heavy oil would delay the tripping time or perhaps not allow the piston to operate at all.

Thermal overload devices (bimetallic type)

1. Indirectly heated bimetallic strip

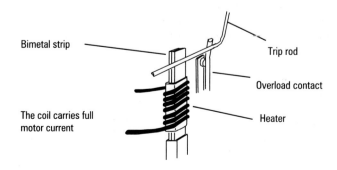

Labels: Bimetal strip, Trip rod, Overload contact, The coil carries full motor current, Heater

Figure 1.49

Principle of operation

The excess motor current causes the heater coil temperature to rise.
The bimetallic strip heats up and bends towards the trip rod.
The trip rod then opens the overload contact at a predetermined point.
The contactor coil de-energises and contactor drops out.

Note: There is one overload unit fitted in each phase line inside the motor starter but just one overload contact.

2. Directly heated bimetallic strip

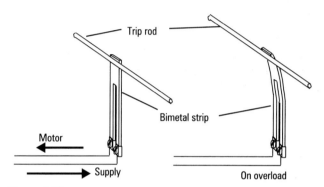

Labels: Trip rod, Bimetal strip, Motor, Supply, On overload

Figure 1.50

The bimetal strip consists of two metals having different coefficients of expansion. When this strip is carrying line current the heating effect of the current causes unequal expansion and bends the strip. This movement is used to move the trip rod and open a contact which de-energises the contactor coil or under voltage release coil causing the contactor to drop out.

Thermistors

A thermistor is a special type of resistor whose resistance varies considerably with changes in temperature.

There are two types of thermistor available:
- the Negative Temperature coefficient (N.T.C.) type, whose resistance decreases as the temperature increases
- the Positive Temperature Coefficient (P.T.C.) type whose resistance increases as the temperature increases.

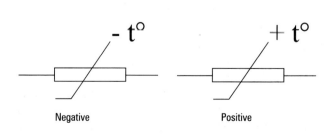

Labels: $-t^{o}$ Negative, $+t^{o}$ Positive

Figure 1.51   Negative and positive thermistor symbols

Overload protective devices

The magnetic_____ trip operates when _____ current causes the solenoid _____ to _____.The plunger opens or closes a _____ in the _____ circuit to de-energise the_____.

Oil _____ time _____ are used to prevent the overload tripping _____ when a motor is drawing _____current whilst being _____.

There are two basic types of bimetallic _____ devices:
1. the _____ heated
2. the_____ heated

The indirectly heated bimetallic _____ bends towards a _____to open a _____when excess _____ causes the _____ temperature to increase.

The _____ strip consists of two _____ having different coefficients of _____.This unequal _____ bends the_____.

The _____ (P.T.C.) type thermistor resistance _____ as the temperature _____.

The _____ (N.T.C.) type thermistor resistance _____ as the temperature _____.

Self-assessment short answer questions

1. (a) Redraw the diagrams shown in Figure 1.52 and show the resultant magnetic field on the diagram.
    (b) State whether the force between the magnets is one of attraction or repulsion.

S	N

N	S

Figure 1.52

2. What do the following symbols represent when applied to an electromagnetic circuit?
    (a) Φ
    (b) F
    (c) B

3. A conductor of 120 mm carrying a current of 10 A lies at right angles to a magnetic field of flux density 5 Tesla. Calculate the force exerted on the conductor.

4. Calculate the magnitude of the induced e.m.f. in a 500-turn coil when the flux linking with the coil changes from 50 mWb to 150 mWb in 150 ms.

5. Briefly explain how an e.m.f. is mutually induced in the secondary winding of a double-wound transformer.

6. (a) Sketch a typical hysteresis characteristic for a ferromagnetic material used in a stator core of an a.c. motor.
    (b) State ONE method of reducing eddy current loss in the stator core.

7. Briefly describe the basic principle of operation of an electromagnetic relay.

8. (a) Draw a labelled sketch of an indirectly heated bimetallic thermal overload device.
    (b) State briefly the principle of operation of the device.

2

Electrostatics

Answer the following questions to remind yourself of what was covered in Chapter 1.

1. Calculate the flux density existing in an area of 15 m^2 if a uniform magnetic flux of 2 Wb exists at right angles to that area.

2. A conductor 0.25 m long is moving through a magnetic field of flux density 0.85 T at a velocity of 30 m/s. Calculate the e.m.f. induced in the conductor.

3. Sketch a typical B/H characteristic for:
   a) iron
   b) air

4. Explain briefly how the resistance of a P.T.C. type thermistor differs from that of a N.T.C. type thermistor with change in temperature.

On completion of this chapter you should be able to:

◆ identify the conditions which give rise to static electricity and describe potential hazards
◆ identify ways of reducing the risks associated with static electrical charge
◆ describe how the plates of a capacitor may be charged
◆ state the units and identify the symbols relevant to capacitance and capacitors
◆ perform calculations relevant to capacitors connected in series and in parallel
◆ identify the construction of different types of capacitor
◆ explain the importance of the working voltage of a capacitor
◆ state the industrial uses of capacitors
◆ explain the reason for connecting discharge resistors to some capacitors

Part 1

Static electricity

All bodies are able to take a "charge" of electricity and this is termed "static electricity". Electricity is "static" if the electric charges do not move as a current in a conductor does but collect in one place (on the plates of a capacitor for example).

Static charges by friction

A static charge can be set up by rubbing certain dissimilar materials against each other. The friction between the materials causes electrons to be transferred from one to the other, resulting in one material being positively charged and the other negatively charged. This effect can be demonstrated by vigorously rubbing a piece of plastic (for example a plastic hair comb) on a piece of wool. Electrons will be transferred from the wool to the plastic causing the comb to become negatively charged. If now the plastic comb is placed near small pieces of paper they will be attracted to the comb and the comb will pick up the pieces of paper.

Note the wool itself is also charged (more positively).

If two pieces of the **same** plastic are rubbed against the same piece of wool they will have like charges in them. Now if one piece is suspended and the other is brought towards it they will repel each other (remember LIKE charges repel).

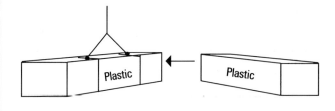

Figure 2.1   Repulsion

Likewise if the suspended piece of plastic is brought towards the piece of wool it will be attracted to the wool (remember UNLIKE charges attract).

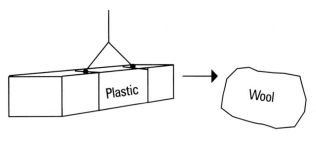

Figure 2.2   Attraction

Remember
A positively charged object attracts a negatively charged object.

Electrostatic discharge

You may have noticed a painful discharge of electricity from your body on touching a metal object after having walked across a nylon carpet.

Touching the door handle of a car even after a short journey can produce a visible discharge of several millimetres.

This is all due to the migration of electrons caused by the movement of insulating surfaces, one upon the other (static charges by friction again). The voltages attained due to such circumstances can be very high and in situations where this can cause problems, steps must be taken to discharge the voltage before any harm can be done.

Persons working with delicate electronic components may have to wear wrist bands which are bonded to remove the charge to earth.

Great care must be taken when handling CMOS (complementary metal oxide semiconductor) devices since they can be damaged by the discharge of static electricity built up on the fingers or particularly nylon shirt cuffs.

Note: CMOS integrated circuits (I.C.s) are supplied in anti-static packages or on special foam pads to reduce the risks associated with static electrical charge.

The process of loading and unloading flammable or explosive substances to and from road vehicles may involve bonding the vehicle to the installation before transferring the load.

Flameproof motors are fitted with metal-alloy fan impellers and must not be replaced with nylon impellers if they become damaged.

Lightning is another form of static electricity in which the cumulative effect of convection over a long period of time causes an enormous build up of electric charge on the underside of a cloud mass which eventually leads to the massive discharge to earth we know as a lightning strike. Lightning discharges can generate large amounts of heat and release considerable forces both due to the large current involved. Lightning conductors are used to protect building structures (for example, multi-storey blocks and towers) to discharge the very high voltage and current safely to earth.

Static electricity was known to man long before cells, batteries and electromagnetic generating devices had been devised.

The word electron is derived from the Greek name for amber which, when rubbed with silk could be made to attract light objects.

Subsequent experiments over the centuries revealed that considerable amounts of electric charge could be accumulated from substances which reacted in the appropriate manner.

The fact that electricity can be stored in static form is of great importance and the effect is usefully employed in all kinds of equipment from minute electronic components to industrial applications of many hundreds of kilowatts.

The capacitor

The capacitor is a device which can store an electric charge. A capacitor usually consists of **two conductive plates,** separated from each other by a layer of **insulation** known as a **dielectric**.

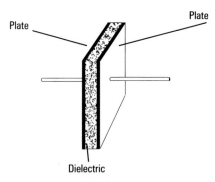

Plate   Plate

Dielectric

Figure 2.3   The basic capacitor

When connected to a source of electromotive force, a capacitor will accept a flow of electric current which will eventually cease when the device is fully charged.

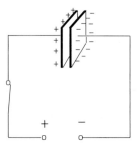

Figure 2.4

Charge

Quantity symbol   Q
Units   coulombs
Abbr.   C

$$Q = It$$

A current of I amperes flowing for t seconds will result in a charge of Q coulombs.

So, for example, when a current of 200 mA flows into a capacitor for 40 mS a charge of 8 mC will be developed.

The amount of charge (the number of coulombs) which can be stored in the capacitor will be determined by the size of the capacitor plates and the properties of the dielectric.

Note: the ideal unit of charge would be that of a single electron, but this inconveniently small. The coulomb is used because it is a much larger unit. (There are approximately 6.3×10^{18} electrons to each coulomb.)

Try this
If a current of 12.5 A flows for a time of 2 minutes determine the amount of charge transferred.

Remember
Current will not "flow" through dielectric materials since they are made from insulating materials. The plates of a capacitor are not in direct contact with each other, so they don't form a circuit in the same way that conductors and resistors do. However, they do have an effect on the current flow in the external circuit to the capacitor. (Refer to apparent current flow in d.c. and a.c. circuits.)

"Apparent" current flow in capacitors – d.c. circuit

In the circuit shown the lamp illuminates for a very short period only, this will be the time taken for the capacitor to "charge" up.

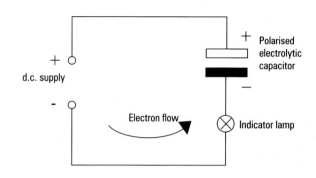

Figure 2.5

Once charged, the electron flow will cease and no more movement takes place. Capacitors can be used to "block" d.c. current. (Figures 2.6 and 2.7)

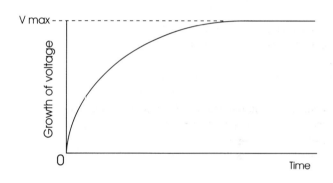

Figure 2.6   *Capacitor charging characteristics (d.c. circuit)*
                       – growth of voltage

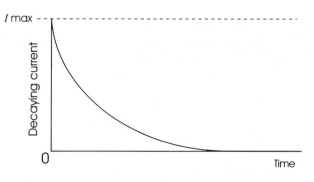

Figure 2.7   *Capacitor charging characteristics (d.c. circuit)*
                       – decaying current

It can clearly be seen from the graphs above that when a capacitor is charged the voltage gradually rises to a maximum (peak) value and at the same time the current immediately rises to a maximum value and gradually falls to zero. So the current is effectively "blocked", since it is zero when the capacitor is charged up to the maximum voltage of the supply.

On discharge a similar pattern of voltage occurs but going
from maximum to zero.

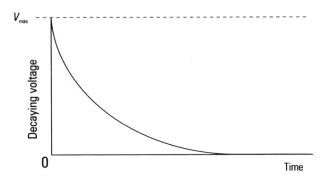

Figure 2.8

"Apparent" current flow in capacitors – a.c. circuit

In the circuit shown the lamp will be illuminated continuously
giving the impression that there is a current flow through the
capacitor. In fact there is not. When a capacitor is connected as
shown in the circuit to an a.c. supply on one half cycle the
capacitor will charge up to a particularly polarity. On the
second half cycle the capacitor will charge up in the reverse
polarity to the first half cycle.

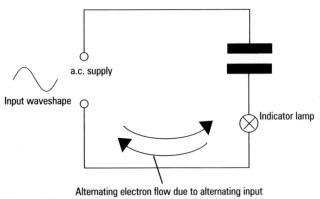

Figure 2.9

The electron movement around the circuit will alternate giving
the impression that there is a current flow through the
capacitor, of course there is not.

The general rule is that capacitors **block d.c.** but **pass a.c.**

Static electricity/charge

All bodies can be electrically _____, this is
termed _____ _____.

When certain dissimilar _____ are vigorously
_____ together a _____ charge can
be set up, this is termed static _____ by
_____.

A positively charge body _____ a _____
charged body.

When working with delicate _____ components
_____ bands can be worn which are
bonded to _____ to safely _____ any
_____ electricity "built up" to
_____.

A capacitor stores an _____ _____, and usually consists of two _____ separated by
a layer of _____ known as the _____.

The amount of charge can be calculated by using the formula
_____.

A capacitor can be used to block a _____
when it is fully _____.

Sketch graphs of
(a) charging current

(b) charging voltage

Part 2

Electric fields and electric flux

Consider the two-plate capacitor in Figure 2.10. When it is charged one of its plates has an excess of electrons and therefore a negative charge, while the other plate has a shortage of electrons and therefore an equal and opposite charge.

These charged plates cause an "electric field" to exist between them in a similar way to that in which a magnetic field exists between the poles of a magnet.

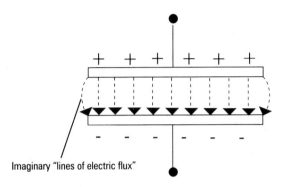

Imaginary "lines of electric flux"

Figure 2.10   *Electric field between charged plates of a capacitor*

The electric field is invisible but can be detected by the effect it has on charged particles placed in it, the particle being attracted to the positive plate if negatively charged, and vice versa. The field is represented by imaginary "lines of electric flux" as shown in Figure 2.10.

Electric flux has the quantity symbol Ψ (pronounced "psi") and has the same unit as charge. Electric flux is therefore measured in coulombs (C), and Ψ is usually replaced by the quantity symbol for charge Q.

Electric field strength
(quantity symbol E, unit V/m)

The intensity of the electric field is called the "electric field strength" and it depends upon:
- the voltage (U) between the plates
- the thickness (d) of the dielectric

It can be calculated using the formula:

$$E = -\frac{U}{d}$$

Note: the minus sign is in the formula because electric fields have a definite direction as well as a magnitude, however, it will be ignored when solving the following problems.

Example

A parallel plate capacitor with a dielectric thickness of 0.5 mm has 1000V applied across it. Calculate the electric field strength in MV/m.

$$E = \frac{U}{d} = \frac{1000}{0.5 \times 10^{-3}}$$

$$= 2 \times 10^6 \text{ V/m}$$

Therefore the electric field strength is 2 MV/m.

Try this

Two parallel plates of a capacitor are separated by a dielectric of thickness 3 mm. Calculate the electric field strength in kV/m when the capacitor is connected to a 150 V supply.

Electric flux density
(quantity symbol D, unit C/m^2)

The electric flux density (D) is defined as the amount of flux per square metre of the electric field. This area (A) is measured at right-angles to the lines of flux. Figure 2.11.

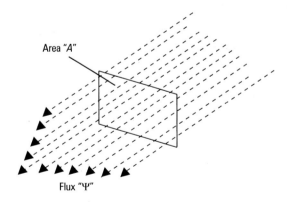

Area "A"

Flux "Ψ"

Figure 2.11   *Cross-section through a uniform electric flux*

The unit of electric flux density is the coulomb per square metre (C/m^2).

Electric flux density can be calculated using the formula

$$D = \frac{\psi}{A}, \text{ but since } \Psi = Q, \text{ then}$$

$$D = \frac{Q}{A}$$

Example
Two parallel plates of a capacitor 50 mm by 25 mm are oppositely charged to a value of 50 mC. Calculate the density of the electric field existing between them.

$$D = \frac{Q}{A} = \frac{50 \times 10^{-3}}{50 \times 25 \times 10^{-6}}$$

$$D = 40 \text{ C/m}^2.$$

Try this
A capacitor is made up of two plates of cross-sectional area 100 cm^2. Calculate the electric flux density when the capacitor is charged to a value of 30 mC.

Remember

	Symbol	Unit
Electric flux	Ψ or Q	coulomb
Electric field strength	E	volts per metre
Electric flux density	D	coulomb per square metre

Capacitance

A capacitor which can store a charge of one coulomb at a potential difference of one Volt is said to have a CAPACITANCE of one FARAD (Abbr. F)

$$C = \frac{Q}{U} \text{ (Farads)}$$

or

$$Q = CU \text{ (Coulombs)}$$

Example
A capacitor of 200 μF which is fully charged at 200 volts holds a charge of

$$Q = CU$$

$$Q = 200 \times 10^{-6} \times 200$$

$$= 40 \text{ mC}$$

Try this
An 80 μF capacitor is fully charged from a 230 V supply. Calculate the charge stored.

The value of capacitance

The value of capacitance is measured in Farads (F) but because of the very large size of this microfarads (μF) are usually used. Occasionally nanofarads and picofarads are used for very small values.

$$1\mu F = 10^{-6}\,F$$
$$1nF = 10^{-9}\,F$$
$$1pF = 10^{-12}\,F$$

These multiples must be used when carrying out calculations.

Table 2.1

Capacitance Conversion Table		
0.000001 μF	= 0.001 nF	= 1 pF
0.00001 μF	= 0.01 nF	= 10 pF
0.0001 μF	= 0.1 nF	= 100 pF
0.001 μF	= 1 nF	= 1000 pF
0.01 μF	= 10 nF	= 10000 pF
0.1 μF	= 100 nF	= 100000 pF
1 μF	= 1000 nF	= 1000000 pF
10 μF	= 10000 nF	= 10000000 pF
100 μF	= 100000 nF	= 100000000 pF

Capacitors in parallel and series

Parallel

When two or more capacitors are connected in parallel, the effect is to increase the plate area under charge and therefore increase the capacitance in direct proportion.

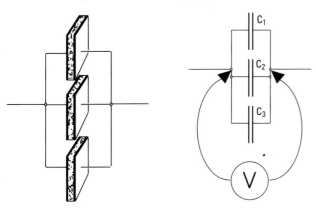

Figure 2.12   Three capacitors connected in parallel

$$C_{(total)} = C_1 + C_2 + C_3$$

Note: the same voltage appears across each capacitor.

Example

A capacitor of 100 μF is connected in parallel with another of 150 μF and then a third of 50 μF. What is the final capacitance?

$$\begin{aligned} C &= C_1 + C_2 + C_3 \\ &= 100 + 150 + 50 \\ &= 300\ \mu F \end{aligned}$$

Try this

Four capacitors are connected in parallel to make up a total capacitance of 68 μF. Two capacitors are 15 μF each. A third is 17 μF. Calculate the value of the fourth.

Because the three capacitors in Figure 2.12 are connected in parallel, each capacitor will take a charge from the supply according to its capacitance, therefore:

$$Q_1 = C_1U$$
$$Q_2 = C_2U$$
$$Q_3 = C_3U$$

but the total charge drawn from the supply must be:

$$Q_{(total)} = Q_1 + Q_2 + Q_3$$

also:

$$Q_{(total)} = C_{(total)}U$$

Example

Two capacitors, of values 20 μF and 60 μF are connected in parallel across a d.c. supply of 100 V. Calculate
(a)   the total capacitance
(b)   the charge on each capacitor
(c)   the total charge

(a) $C_T = C_1 + C_2 = 20 + 60 = 80\ \mu F$
(b) $Q_1 = C_1U = 20 \times 10^{-6} \times 100 = 0.002\ C$
      $Q_2 = C_2U = 60 \times 10^{-6} \times 100 = 0.006\ C$
(c) $Q_T = Q_1 + Q_2 = 0.002 + 0.006 = 0.008\ C$
      or $0.008 \times 10^3 = 8\ mC$

Series

When capacitors are connected in series, the effect on overall capacitance is quite different.

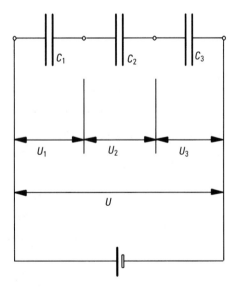

Figure 2.13

The charging current is the same in all three capacitors and flows for the same length of time because the three capacitors are connected in series.

i.e. $\qquad Q_1 = Q_2 = Q_3$

Indeed it can be said that the charge in the whole network is equal to that of any individual capacitor because the charging current in the whole series circuit is the same as any individual capacitor.

Then $\qquad Q_{(total)} = Q_1 = Q_2 = Q_3$

It is obvious that the total voltage is the sum of the individual capacitor voltages

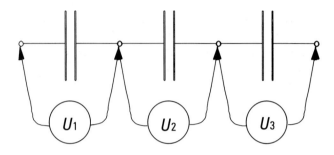

Figure 2.14

$$U_{(total)} = U_1 + U_2 + U_3$$

but $\qquad U = \dfrac{Q}{C}$ (by transposition)

therefore $\dfrac{Q}{C_{(total)}} = \dfrac{Q}{C_1} + \dfrac{Q}{C_2} + \dfrac{Q}{C_3}$

As Q is the same in each case, then we can divide each term by Q, (in layman's terms the Qs cancel out)

and this gives us;

$$\frac{1}{C_{(total)}} = \frac{1}{C_1} + \frac{1}{C_2} + \frac{1}{C_3}$$

to take this a stage further

$$; \qquad C_{(total)} = \frac{1}{\dfrac{1}{C_1} + \dfrac{1}{C_2} + \dfrac{1}{C_3}}$$

Example
A series circuit consists of three capacitors of different capacitances as follows;

$$C_1 = 6 \text{ μF}$$
$$C_2 = 3 \text{ μF}$$
$$C_3 = 2 \text{ μF}$$

What is their combined capacitanc

$e?\ C_{(total)} = \dfrac{1}{\dfrac{1}{6} + \dfrac{1}{3} + \dfrac{1}{2}}$

$= 1\ \mu F$

It may be easier to think of combined capacitor circuits as being similar to, yet differing from, resistor networks as follows.

1. The combined value of several resistors in **SERIES** can be found by:

   $R_t = R_1 + R_2 + R_3$ etc

   The combined value of several capacitances in **PARALLEL** can be found by:

   $C_T = C_1 + C_2 + C_3$ etc

Note: If only two capacitors are connected in series the total capacitance may be obtained by using the "product over sum" method:

$$C_T = \frac{C_1 \times C_2}{C_1 + C_2}$$

Example

An 8 μF and a 12 μF capacitor are connected in series across a 150 V d.c. supply.

Calculate:

(a) the total capacitance

(b) the charge on each capacitor

(a) $C_T = \dfrac{C_1 \times C_2}{C_1 + C_2} = \dfrac{8 \times 12}{8 + 12} = \dfrac{96}{20} = 4.8\ \mu F$

(b) $Q_T = C_T\,U = 4.8 \times 10^{-6} \times 150 = 0.00072\text{C}$
   or $0.00072 \times 10^6 = 720\ \mu C$
   and since $Q_T = Q_1 = Q_2$ the charge on each capacitor
   $= 720\ \mu C)$

Try this

Two 10 μF capacitors are connected in series across a 100 V d.c. supply.

Determine

(a) the equivalent capacitance of the two

(b) the total charge

(c) the charge on each capacitor

---

2. The combined value of several resistors in **PARALLEL** can be found by

$$R_T = \frac{1}{\dfrac{1}{R_1} + \dfrac{1}{R_2} + \dfrac{1}{R_3}} \quad \text{etc.}$$

The combined value of several capacitances in **SERIES** can be found by:

$$C_T = \frac{1}{\dfrac{1}{C_1} + \dfrac{1}{C_2} + \dfrac{1}{C_3}} \quad \text{etc.}$$

Try this

1. Four capacitors are connected in series. Their values are;

   $C_1 = 200\,\mu F$
   $C_2 = 400\,\mu F$
   $C_3 = 600\,\mu F$
   $C_4 = 800\,\mu F$

   What is their combined capacitance?

2. What is the value of this arrangement?

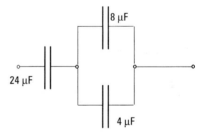

Figure 2.15

The energy stored in a capacitor

If a capacitor is charged at a constant rate of current the potential difference between the plates will increase at a constant rate.

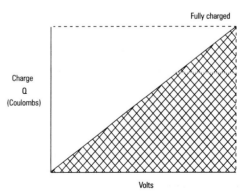

Figure 2.16   Voltage charge characteristic

The charge is in coulombs which is ampere seconds.

The potential difference is in volts.

Multiplying these two quantities together gives

$$U \times I \times t \text{ (watt seconds)}$$
$$\text{or joules (of energy)}$$

But $\qquad Q = CU$

and if this is multiplied by the volts will give

$$CU^2$$

which is also an expression of energy in Joules.

As you can see, this graph, (Figure 2.16) because it has a constant charging current, is a straight line which exactly bisects the figure and which has an area $.CU \times U$

Therefore; the energy stored in a capacitor can be found by

$$W = \frac{1}{2}CU^2 \text{ Joules}$$

Example

What is the energy stored in a 160 µF capacitor if the p.d. between the plates is 400V?

$$W = \frac{1}{2}CU^2$$

$$= \frac{1}{2} \times 160 \times 10^{-6} \times 400^2$$

$$= 12.8 \text{ Joules}$$

Try this

How much energy is stored in a 64 µF capacitor at a potential difference of 100V?

Points to remember ◀ ─ ─ ─ ─ ─ ─ ─ ─

Electric fields/capacitance in series and parallel/energy stored

The electric field is represented by imaginary lines of
_____ _____.

Electric field strength has the quantity symbol _____ and is measured in _____.

D is the quantity symbol for _____ _____ ____ and its unit is _____.

Show the following in farads
1 µF = _____ 1 nF = _____ 1 pF = _____

Three capacitors in parallel can be calculated using the formula:

The formula $\dfrac{1}{C_{(total)}} = \dfrac{1}{C_1} + \dfrac{1}{C_2} + \dfrac{1}{C_3}$ is used to determine

_____.

When only two capacitors are connected in _____ the "product over sum" method can be used, which is

$$C_T =$$

In a series circuit the charge on each capacitor is the _____.

$W = \dfrac{1}{2}CU^2$ is used to determine the_____ and

is measured in _____.

Part 3

Construction of capacitors

There are very few electronic circuits that do not contain a capacitor of some type or other. They can be fixed or variable and range in size from small plastic blobs the size of a solder drip to large cans that look as though they could hold food.

Figure 2.17   Sub-miniature ceramic capacitor

Figure 2.18   General purpose electrolytic capacitor

Figure 2.19   Capacitor: general symbol

General construction of a cylindrical (tubular) type capacitor

This type of capacitor normally consists of two strips of aluminium foil, separated from each other by strips of paper, rolled up and inserted in a plastic or metal cylinder.

Plate area however is not the only factor which determines the capacitance of the device. There are three factors which affect capacitance;
1. The surface area of the plates
2. The distance between the plates
3. The insulating material between the plates, which is also known as the dielectric.

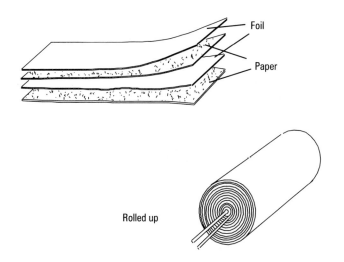

Figure 2.20   Foil type capacitor

1. The foil strips form the plates of the capacitor and consequently the larger the plate area, the greater the capacitance. Using very thin foil allows large plate areas to be accommodated in relatively small enclosures.
    In some types of capacitor, multiple plates are connected in parallel groups to increase the overall area.
2. For maximum capacitance, the plates must be as close as possible. For this reason, the insulating layer is kept very thin. A few thousandths of a millimetre of variation in the thickness of the dielectric can have a significant effect on the capacitance.
3. The dielectric has to resist the flow of electrons, sometimes at very high voltages and it is a matter of choice which material is to be used. Some types of capacitor have an air gap between the plates because air is quite effective as a dielectric but as can be seen from Table 2.2 there are other materials which are better.

Table 2.2

Material	Dielectric constant
Air	1.0
Aluminium oxide	10.0
Glass	7.6
Mica	7.5
Mylar	3.0
Paper	2.5
Porcelain	6.3
Quartz	5.0
Tantalum oxide	11.0
vacuum	1.0

Types of capacitors

To a large extent the application determines the type of capacitor that can be used. As with resistors there are fixed and variable types of capacitor.

Fixed capacitors can be placed into three general classes related to their dielectrics.

Low loss, high stability
- mica
- low K ceramic
- polystyrene

Medium Loss, medium stability
- paper
- plastic film
- high K ceramic

Polarised capacitors
- electrolytic
- tantalum

Details of the different types of capacitor can be found in "Basic Science and Electronics" one of the books in the Foundation Course.

Polarity

Once the size, type and voltage rating of a capacitor has been decided it only remains to ensure that its polarity is known.

As we have seen, some capacitors are constructed in such a way that if the device is operated with the wrong polarity its properties as a capacitor will be destroyed.

This is particularly true of electrolytic types.

The positive terminal must never be allowed to go negative whether by wrong connection, supply reversal or by connection to an alternating voltage.

Polarity may be indicated by a + or – as appropriate. Electrolytics contained in metal cans may use the can as a negative connection. If no other marking is indicated but if it is still suspected that the capacitor is polarised a slight indentation in the case will indicate the positive end.

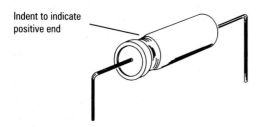

Indent to indicate positive end

Figure 2.21

Tantalum capacitors having a spot on one side are polarised as shown in Figure 2.22. When the spot is facing you the right hand lead is positive.

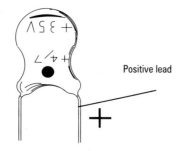

Positive lead

Figure 2.22

Suppression capacitors

A number of capacitors can be connected inside a single container to make up a suppression capacitor.

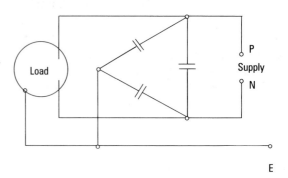

Load

P
Supply
N

E

Figure 2.23   Suppression capacitor circuit

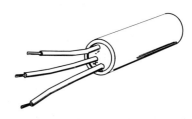

Figure 2.24   Suppressor capacitor

Points to remember ◀ – – – – – – – – – – – – –

Capacitors/construction/types
The three main factors which affect capacitance are:

1.

2.

3.

Fixed capacitors can be placed into three general classes:

1._____, high stability

2._____, medium stability

3._____

A polarised capacitor is one in which the dielectric is formed by passing a _____ through it.

If a polarised capacitor is connected the wrong way round it will _____.

The circuit symbol for a variable capacitor is _____ and for a pre-set variable is _____.

Part 4

Dielectric strength and working voltage

A dielectric material, like any insulating material, will break down when subjected to a higher voltage than it is designed for. When an insulator breaks down its resistance suddenly falls to a low value, consequently it will pass more current than normal.

Note: all insulators pass some "leakage current" but this is normally very small.

Dielectric strength relates to the voltage that a dielectric can withstand without breaking down, and it is measured in volts per metre (V/m). The thicker the dielectric material is, the greater is its ability to withstand breakdown.

The maximum voltage that a capacitor can withstand is known as its **"maximum working voltage"** which is normally marked on the side of a capacitor as well as its capacitance. The voltage may be marked as "d.c. wkg" (d.c. working voltage) or as "a.c. r.m.s." (root-mean-square working voltage). For example, a capacitor marked as shown in Figure 2.25 below has a maximum d.c. working voltage of 400V and may be used at any voltage up to this value, and when used on a.c. the peak voltage should not exceed the d.c. rating of the capacitor.

Figure 2.25

Remember
On a.c. circuits 230 V is the r.m.s. value of the supply voltage and the maximum (peak) voltage

$$= \frac{230}{0.707} = 325 \text{ V approx.}$$

So a capacitor with a maximum working voltage above this value would be preferable (for example a 400 V capacitor would be suitable).

The use of capacitors

Details on the use of capacitors can be found in the Foundation book "Basic Science and Electronics". Two other uses not included in that publication are:

1. Power factor correction capacitors handling many hundreds of amperes are to be found in industrial installations where a lagging power factor, caused by heavy motor loads would otherwise cost the consumer dearly in tariff surcharges.
2. The single phase capacitor-start motor needs the phase shifting effect of the capacitor to make it rotate.

Try this
A series circuit consists of four capacitors of 20 μF, 40 μF, 80 μF and 100 μF respectively connected to a 440 V supply.

Calculate:
(a) single capacitance required to replace the four
(b) total energy stored across all four capacitors

Discharge resistors

It is not uncommon to find a capacitor with a resistor connected across its terminals. The reason for this is that once a capacitor is charged, it can hold its charge for a considerable length of time. Any person coming into contact with the capacitor could receive a painful electric shock long after the mains supply has been isolated.

Accidents have resulted from persons receiving an electric shock from capacitors when working on ladders or scaffolds causing them to step back or lose their grip and suffer injury from falling.

A discharge resistor fitted to the capacitor will safely dissipate the charge and help to prevent accidents and injuries.

The choice of discharge resistor will depend on the circumstances, but it must have a low enough value to discharge the capacitor quickly enough to prevent inadvertent electric shock. At the same time it must not be so low as to interfere with the correct operation of the circuit.

It must be remembered that any two conductive surfaces which are separated from each other by an insulating medium can form a capacitor.

It is frequently found that the capacitance between the cores of a multicore cable can produce unexpected effects, particularly when low values of current and voltage are being used.

Where long cables are used for control circuitry in an a.c. system care should be taken to ensure that alternating current does not continue to flow due to the capacitance of the cable when the circuit has been opened.

Another situation from which danger can arise is after the isolation of long overhead conductors. The overhead cable can and will hold a charge, sometimes at a very high voltage between the conductors and earth.

When isolating such a system it is proper procedure to connect the conductors to earth and keep them earthed while work is being carried out. Most H.V. circuit breakers have an earthing facility, i.e. ON – OFF – EARTH, which will ensure safety during maintenance operations.

It would not be considered good practice for anyone, however brave, to approach a high voltage overhead cable with an earthed "croc clip".

Safety

Be careful when handling large capacitors for they may still be charged long after the supply has been isolated.

Points to remember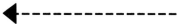

Dielectric strength/working voltage/use of capacitors/discharge resistors
A dielectric material will break down when it is subjected to a _____ than it can withstand.

The maximum voltage that a capacitor can withstand is known as its _____.

A timing circuit consists of a _____ arrangement.

The capacitor connected across the supply to a fluorescent luminaire is for _____, and the one in the starter switch suppresses _____.

A smoothing capacitor is used to reduce _____.

A discharge resistor is fitted across a capacitor's _____ to safely dissipate the _____.

Self assessment short-answer questions
1. State three factors which affect the capacitance of a capacitor.

2. Name three types of dielectric material used in the construction of capacitors.

3. Calculate the equivalent capacitance for the circuit shown in Figure 2.26.

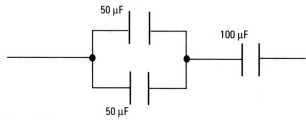

Figure 2.26

4. State the unit for each of the following:
   (a) electric field strength
   (b) electric flux density
   (c) charge on a capacitor

5. Explain why:
   (a) a capacitor is fitted inside a fluorescent luminaire across the supply terminals
   (b) it is important to choose a capacitor with the correct working voltage
   (c) discharge resistors are commonly connected to large capacitors

6. (a) Identify ONE potential hazard that could be created by static electricity.
   (b) Explain ONE method of reducing the hazards that may be caused by static electricity.

3

D.C. machines

Answer the following questions to remind yourself of what was covered in Chapter 2.
1. Convert the following into farads:
   (a) 2 μF
   (b) 4 nF
   (c) 8 pF

2. Capacitors of capacitance 4 μF, 6 μF and 12 μF, respectively, are connected in series to a 100 V d.c. supply.
   Calculate:
   (a) the total capacitance
   (b) the total charge
   (c) the charge on each capacitor

3. What is the energy stored in a 100 μF capacitor if the p.d. between the plates is 100 V?

On completion of this chapter you should be able to:

- describe with the aid of diagrams the construction of d.c. machines
- state that most d.c. machines will operate as a motor or as a generator
- explain the operation of a.c. motors and generators
- draw simple circuit diagrams for d.c. machines
- describe the operating characteristics of d.c. motors and generators

Part 1

D.C. motors and generators

The actual construction of a motor and a generator is basically the same. It follows then that when a motor is powered from a supply and is running at full speed it will also be generating a voltage. As it can be determined from Lenz's Law, this induced e.m.f. is opposite in direction to the supply voltage. It is known as the "back e.m.f.". This generated voltage is always less than the supply due to the voltage drop in the armature windings. The voltage drop in these windings can be calculated using Ohms Law.

Armature voltage drop
$$= \text{armature current} \times \text{armature resistance}$$

$$= I_a \times R_a$$

It can also be shown that

Armature voltage drop
$$= \text{supply voltage} - \text{back e.m.f.}$$

$$I_a R_a = U - E$$

where

U = supply voltage
E = back e.m.f.
I_a = armature current
R_a = armature resistance

To calculate the supply voltage for a given motor the formula can be transposed so that

$$U = E + I_a R_a$$

Example
When connected to a 200 V supply a d.c. motor has a current of 0.25 A flowing through the armature which has a resistance of 20 ohms. Calculate the back e.m.f. generated in the motor.

First the formula must be transposed so that the back e.m.f. Is the subject. This can be carried out from $I_a R_a = U - E$

So then $E = U - I_a R_a$

Filling in the details

$$E = 200 - (0.25 \times 20)$$
$$= 200 - 5$$
$$= 195 \text{ V}$$

Remember

Lenz's Law
The direction of an induced e.m.f. is always such that it sets up a current opposing the change of flux responsible for inducing that e.m.f.

Try this

1. A d.c. motor has an armature which has a voltage drop of 7.5 V when connected to a 250 V d.c. supply. Calculate the back e.m.f. of the motor.

2. An armature in a d.c. motor has a resistance of 0.15 ohms and creates a voltage drop of 6.25 V when the machine is connected to a 230 V d.c. supply. Calculate
   (a) the back e.m.f of the motor
   (b) the current flowing in the armature

The back e.m.f. of a motor is also proportional to the magnetic flux of the poles and the speed of rotation. This can be expressed as

$$E \propto \Phi N$$

where

Φ = flux/pole in weber (Wb)
N = speed in rev/min

This can be transposed for speed so that

$$N \propto \frac{E}{\Phi}$$

Later on, when speed control is considered, the fact that the speed is inversely proportional to the flux will prove very useful.

Generators produce an output voltage due to the movement of the armature conductors through the magnetic flux. As the armature conductors all carry current and have a resistance, it follows that they also produce a voltage drop. The actual terminal voltage of a generator is always slightly less than the generated voltage due to this voltage drop in the armature. This can be expressed as

$$U = E - I_a R_a$$

where

U = terminal voltage (V)
E = induced (generated) e.m.f. (V)
I_a = armature current (A)
R_a = armature resistance (ohms)

Example
What is the terminal voltage of a d.c. generator which has an armature resistance of 0.64 ohms and a current flowing through it of 3.25 A when it generates 250 V?

$$U = E - I_a R_a$$

filling in the details

$$U = 250 - (3.25 \times 0.64)$$
$$= 247.92 \text{V}$$

Remember
Most d.c. machines will operate as a motor or as a generator.

1. A d.c. generator supplies 340 V when the armature current is 10.5 A and the resistance is 0.45 ohms. Calculate the generated e.m.f.

2. The generated voltage of a d.c. generator is 230 V and it has an armature resistance of 0.34 ohms. Calculate the terminal supply voltage when the armature current is
    (a) 15 A

    (b) 25 A

    (c) 50 A

Construction of a d.c. motor

Figure 3.1 shows the main components of a typical d.c. motor which is used to start a car.

The field

The magnetic field in most motors is produced by electromagnets, unlike the motors in toy trains and cars that use permanent magnets. The magnetic field produced in the field windings has to be strong enough to go into the armature and interact with those windings. To ensure the strength of field is kept as strong as possible it must be carried in good magnetic conductors. The poles and case are constructed of iron which conducts the magnetic flux to the required parts of the motor. There will always be an air gap between the poles and the armature to allow it to rotate. As air is not a good conductor of magnetism the gap must be kept to a minimum. When current flows in the field windings a complete magnetic system of circuits is produced. Figure 3.2 shows typical magnetic circuits for a four pole motor.

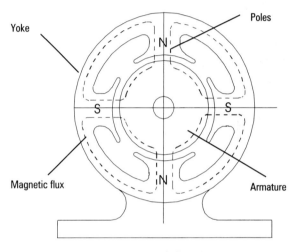

Figure 3.2   Magnetic circuit for a four pole motor

The magnetic flux flows in complete circuits going through one pole, round the frame, which when used in this way becomes the yoke, through the other pole and across to the first pole through the steel core of the armature to complete the magnetic circuit. The only air gaps in this circuit are between the armature and the pole shoes.

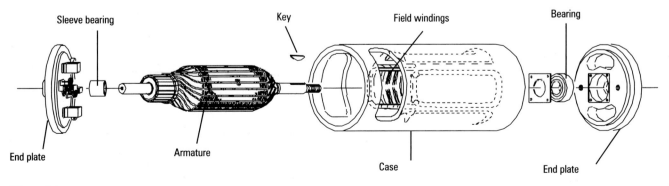

Figure 3.1

The poles on small motors are often made in one piece, with the core and shoe being shaped to take the winding. On larger motors the shoe is bolted to the pole core after the field coils are fitted. In these motors the pole core is usually made of solid iron and the pole shoe is laminated to reduce the effects of eddy currents being produced by the rotating armature conductors. The field windings are coils preformed to a shape so that they fit onto the pole core.

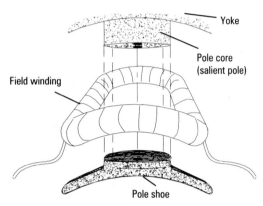

Figure 3.3   A complete pole with winding

The armature

The armature is made up of high quality magnetic alloy laminations mounted on a steel shaft with a commutator assembly at one end. The commutator consists of the appropriate number of copper segments all insulated from each other by mica based compositions. Each segment is dovetailed into an insulating ring to prevent movement due to heat and centrifugal forces. The whole commutator assembly is keyed onto the shaft.

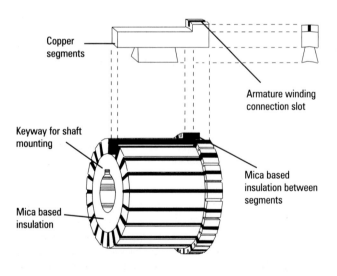

Figure 3.4   A commutator assembly with one segment removed to show the make up

The armature coils are usually preformed to the required shape before assembly. The coils are then fitted into the slots in the laminations and the ends terminated into the commutator connection lugs. The method used to insulate the windings from the laminations depends on the size and type of motor. This may be just a type of varnish on small motors or impregnated cloth on larger ones.

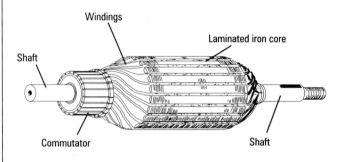

Figure 3.5   Complete armature assembly

After the armature is completely assembled and all of the windings are securely fitted, it must be checked to ensure it is electrically and mechanically sound. Insulation tests at the appropriate voltages must be carried out and the results recorded. The whole assembly must also be checked to ensure it is in alignment and balanced.

Brush connections

Electrical connection is made to the commutator through carbon or graphite brushes. Each brush is held in place by a box arrangement which is fitted securely to the motor frame. The pressure that ensures good electrical contact is produced by a flat spring pushing onto the top of each brush. As the brush wears down through use, the spring is designed to keep a constant pressure. The brush is connected to the electrical circuit by a flexible "pigtail" fitted into the top of the brush and terminated onto the windings away from any movement.

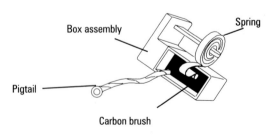

Figure 3.6   A brush in a box

Construction/operation of d.c. machines

The induced e.m.f. in a d.c. motor is in an _____ direction to the _____ voltage and is known as the _____ e.m.f.

The supply voltage to a d.c. motor can be calculated using the formula:

The speed of a d.c. motor varies directly with its _____ and inversely with its _____.

The terminal voltage of a generator can be calculated using the formula

$$U = E - I_a R_a$$

Where
- U =
- E =
- I_a =
- R_a =

The frame of a d.c. machine is known as the _____.

The magnetic _____ in most d.c. motors is produced by _____ and in very small motors it is produced by _____.

The armature coils are fitted into _____ in the laminated _____.

Part 2

D.C. motors

There are three main types of d.c. motor, series, shunt and compound wound

The series motor

The first motor to consider in some detail is the series type. As the name implies, the field and armature windings are all connected in series as shown in Figure 3.7.

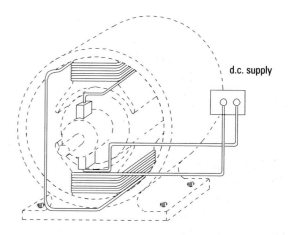

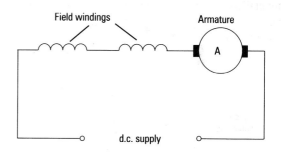

Figure 3.7   *A connection diagram, and circuit diagram, for a series wound motor*

In a series motor the current flowing in the armature also flows through the field windings, so when the motor is put under load, the armature and field currents increase. This means that the magnetic flux becomes stronger and more torque is produced. When the load is reduced the magnetic flux becomes weaker and the speed increases. In theory, if a series motor was left with no load connected it would continue to increase speed until it destroyed itself.

This type of motor is used where large starting torques are required, such as on traction engines and cranes. Figures 3.8a and b show the relationships between speed/torque, and torque/load, for this type of motor.

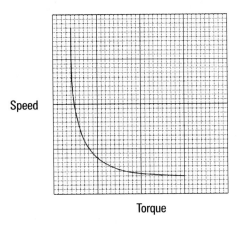

Figure 3.8a

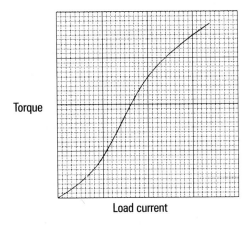

Figure 3.8b

Speed control of a series motor

The speed on this type of motor can be controlled by using "diverter" resistors across either the field or the armature windings. The field diverter shunts off some of the current from the field windings making the field weaker and therefore increasing the speed. When a diverter resistor is used across the armature the current in the field windings is increased and the motor is slowed down.

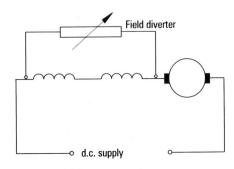

Figure 3.9   Field diverter circuit

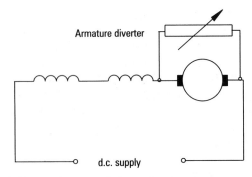

Figure 3.10   Armature diverter circuit

Starting

Starting a d.c. series motor must be carried out using the correct equipment. When the motor is stationary no back e.m.f. is being produced so it will take the maximum current from the supply. To reduce this current while the motor is building up its speed and its back e.m.f., a number of resistors are connected in series with it. These reduce the supply voltage and hence the current drawn by the motor. As the speed increases the resistors are shorted out so that when the motor reaches full speed it is connected straight across the supply. At this stage the back e.m.f. is at maximum so the resistors are no longer required.

Example

Assume that a d.c. motor with an armature resistance of 0.1 Ω is switched directly, without a starter, on to a 200V d.c. supply.

$$U = E + I_a R_a$$

$$\therefore \quad I_a = \frac{U - E}{R_a}$$

At the instant of switching on the motor will be stationary, so the back e.m.f. will be zero.

$$\therefore \quad I_a = \frac{U - 0}{R_a} = \frac{200 - 0}{0.1} = 2000 \text{ A}$$

Try this
A 240 V d.c. motor has an armature resistance of 0.2 Ω, calculate the starting current if no starting resistor is connected into the armature circuit. Ignore the field resistance.

The starter used for this is also designed to give overcurrent protection and "no volt" release. Overcurrent protection is monitored by a coil connected in series with the motor, so that if the supply current increases beyond a predetermined value, the coil becomes magnetically energised. At this stage an armature is attracted to the coil and a pair of contacts are shorted out. These contacts are connected across the electromagnet that holds the starting arm in the run position. When the electromagnet becomes de-energised the starting arm automatically returns to the "off" position. This same electromagnet also acts as the "no volt" protection, so should there be a power failure the machine will automatically switch off until it is manually restarted. This type of starter is often referred to as a "faceplate starter" due to its flat vertically mounted construction. Figure 3.11 shows a circuit diagram for a faceplate starter for a series motor.

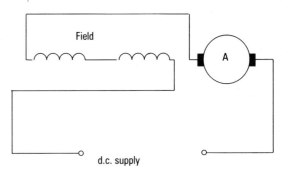

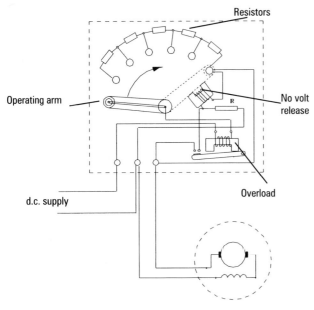

Figure 3.11   Faceplate starter for a series motor

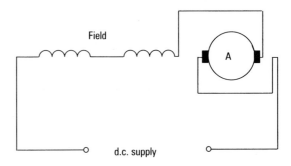

Figure 3.12   Changing the direction of rotation of a d.c. series motor

Reversing the direction of rotation

In Chapter 1, it was seen that when a single current-carrying conductor was placed in a magnetic field and the direction of the current was reversed then the direction of the movement in the conductor was also reversed. However, when BOTH the magnetic field and current were reversed then the movement on the conductor remained the same as before the changes were made. D.C. motors respond in a similar way. If the supply polarity is changed the motor will still rotate in the same direction as before. This is because in effect both the armature and field windings have been reversed. If either the field or the armature connections are changed over the motor will rotate in the opposite direction, but if both are changed the direction of rotation stays the same. Figure 3.12 shows the circuit diagrams for changing either the field or armature connections. Refer to Figure 3.7 for the original connections and then compare it with the two circuits in Figure 3.12.

Remember

Series motor

Windings
  Armature and field windings connected in series.

Load type
  On high load speed decreases and torque increases.

Starting
  Variable resistances connected in series.

Speed control
  By variable resistors connected either across the armature or field windings.

Reversing rotation
  Changing either the armature or field connections but NOT both.

The shunt motor

This is a motor where the field and armature windings are connected in parallel, as shown in Figure 3.13

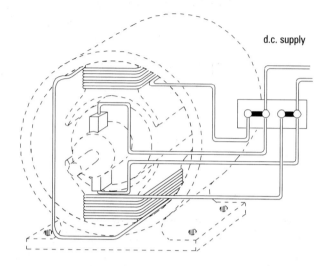

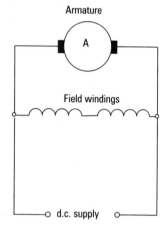

Figure 3.13   Connection diagram and circuit diagram of a shunt wound motor

When a motor is connected in this way the field windings receive the full supply voltage across them. As the supply is constant the field strength is constant and therefore the motor speed is fairly constant. As the field strength does not vary Φ is constant and the speed (n) is proportional to the back e.m.f. (E). As the load is increased the armature current and voltage drop also increase producing a slight reduction in motor speed. Figures 3.14 a and b show the characteristics for this type of motor.

This type of motor is used where virtually constant speed is required on drives such as machine tools, fans and conveyor systems.

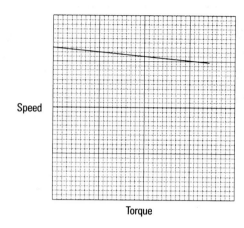

Figure 3.14a

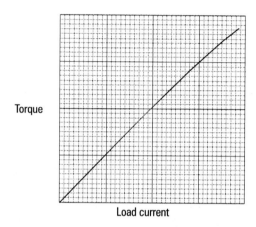

Figure 3.14b

Speed control of a shunt motor

Speed control of shunt motors is obtained by connecting a variable resistor in series with the field windings, as shown in Figure 3.15. When this is adjusted to increase the resistance, the current is reduced and so the magnetic field strength is weakened. To produce the same back e.m.f. the motor has to run faster so the speed increases. Shunt motors can have their speed varied using resistors in this way within a range of about 3 to 1 without weakening the poles too much.

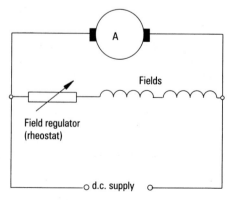

Figure 3.15   Circuit diagram of shunt motor with field regulator

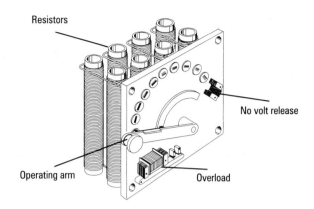

Figure 3.17   *Construction of a faceplate starter*

Starting

The starting procedure for a shunt connected motor is similar to that of the series motor. In both types a faceplate starter is used to add resistance into the circuit during the starting process. When starting the shunt connected motor the resistances are connected into the armature circuit only as this is where the high starting currents will be. When the armature is rotating at full speed and the maximum back e.m.f. is being produced, the resistors are shorted out. Overcurrent and no voltage protection are similar to those on the series type starter. Figure 3.16 shows a circuit diagram for a faceplate starter for a shunt connected motor.

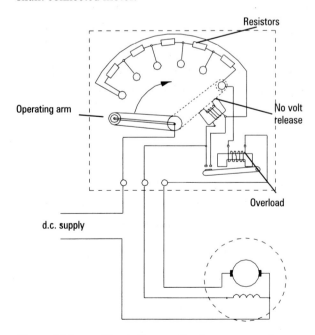

Figure 3.16   *Circuit diagram for shunt connected motor*

Reversing the direction of rotation

Like the series motor the shunt connected motor has to have either the connections to the field windings or armature windings changed over, not both. Figure 3.18 shows circuit diagrams of how this can be carried out. See Figure 3.15 for the original circuit.

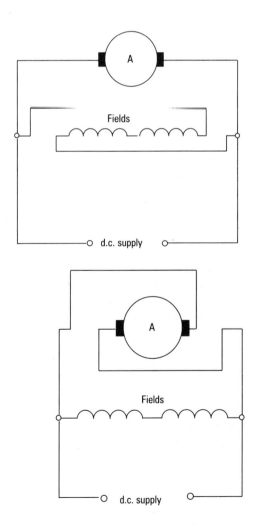

Figure 3.18   *Changing the direction of rotation of a d.c. shunt motor*

Shunt motor

Windings
> Armature and field windings are connected in parallel across the supply.

Load type
> As the load increases the torque increases and the speed slightly decreases.

Starting
> Variable resistances are connected into the armature circuit.

Speed control
> By using variable resistors connected in series with the field windings.

Reversing rotation
> Changing either the armature or field connections but NOT both.

The compound motor

This type of motor has both shunt and series connected field windings, as shown in Figures 3.19, 3.20 and 3.21.

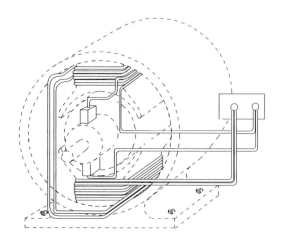

Figure 3.19   Compound connected motor

The characteristics of the motor depend on which set of field windings is the strongest. If the series winding is used to increase the magnetic pole strength as the load current increases, this is known as cumulative compounding. This type of motor has characteristics between the series and shunt types, giving high starting torques with safe no-load speeds. This makes cumulative compounded motors ideal for heavy intermittent loads such as lifts and hoists.

There are two methods of connecting the shunt windings of a compound motor. Figure 3.20 shows the long shunt connections, whereas Figure 3.21 is connected in short shunt.

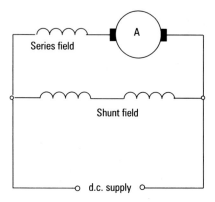

Figure 3.20   Long shunt connected

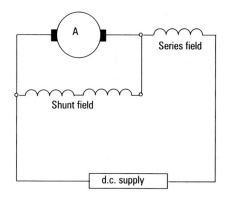

Figure 3.21   Short shunt connections

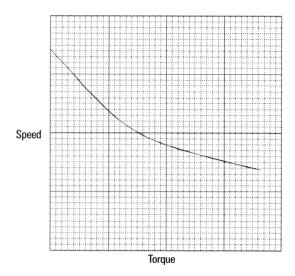

Figure 3.22   Characteristics of a cumulative compounded motor

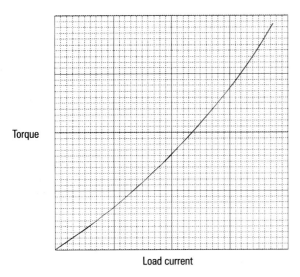

Torque

Load current

Figure 3.23   *Characteristics of a cumulative compounded motor*

Speed control of a compound motor

When the series winding is used to weaken the magnetic pole strength as the load current increases, this is a differentially compounded motor. In these the magnetic flux produced by the series field winding opposes and has the effect of weakening the magnetic flux set up by the parallel field winding, which then increases the motor speed as load current increases. These motors have a tendency to become unstable as the differential compound effect can cause the speed of the motor to rise to uncontrollable levels. This instability means that this type of machine is seldom used in practice. It is important to know what can result so that motors are not accidentally connected in this way.

The speed of cumulative compound motors can be controlled by either series or shunt connected variable resistors. In some circumstances both series and shunt are used.

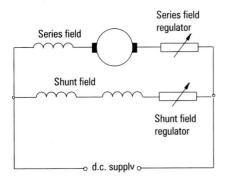

Figure 3.24   *Circuit diagram of compound motor with both series and shunt connected variable resistors*

Reversing the direction of rotation

To reverse the compound motor the connections to the armature are changed as shown in Figure 3.25.

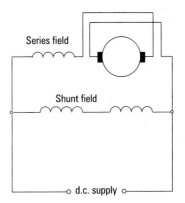

Figure 3.25

To reverse the direction using the field windings, both sets have to be reversed as shown in Figure 3.26.

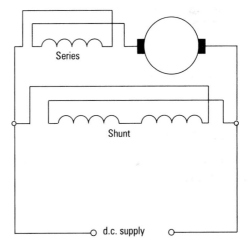

Figure 3.26

Note: If **only one field** is reversed the characteristics of the motors will be changed. (i.e. wrong connection of the fields will alter the fields from cumulative to differential or vice versa.)

Remember

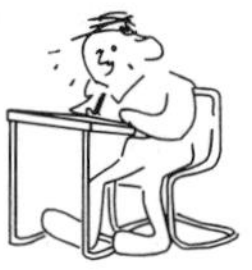

Compound motor

Windings

Part of the field is in series with the armature and part is in parallel across the supply.

Load type

As the load increases the speed decreases and the torque increases.

Starting

Variable resistors are used to reduce the starting current.

Speed control

By variable resistors in the armature and/or field windings.

Reversing rotation

Changing either the armature or both sets of field connections but NOT both.

Points to remember

D.C. motor circuits/operating characteristics

When the load on a d.c. series motor is reduced the magnetic flux becomes ________________ and the speed ______________.

The speed of a d.c. series motor may be controlled by connecting ______________ resistors across the ______________ or the ______________.
The armature and the ______________ of a d.c. shunt motor are connected in ______________.
The faceplate starter is used to add ______________ into the ______________circuit during the ______________ process of a d.c. motor to limit the ______________.

The direction of rotation of a shunt wound motor may be reversed by changing either the______________ or ______________connections but ______________ both.

A compound motor has a ______________ and a ______________ field winding.

There are two ways of connecting a compound motor and they are

- a ______________ ______________ connection
- a ______________ ______________ connection

Part 3

Construction of a d.c. generator

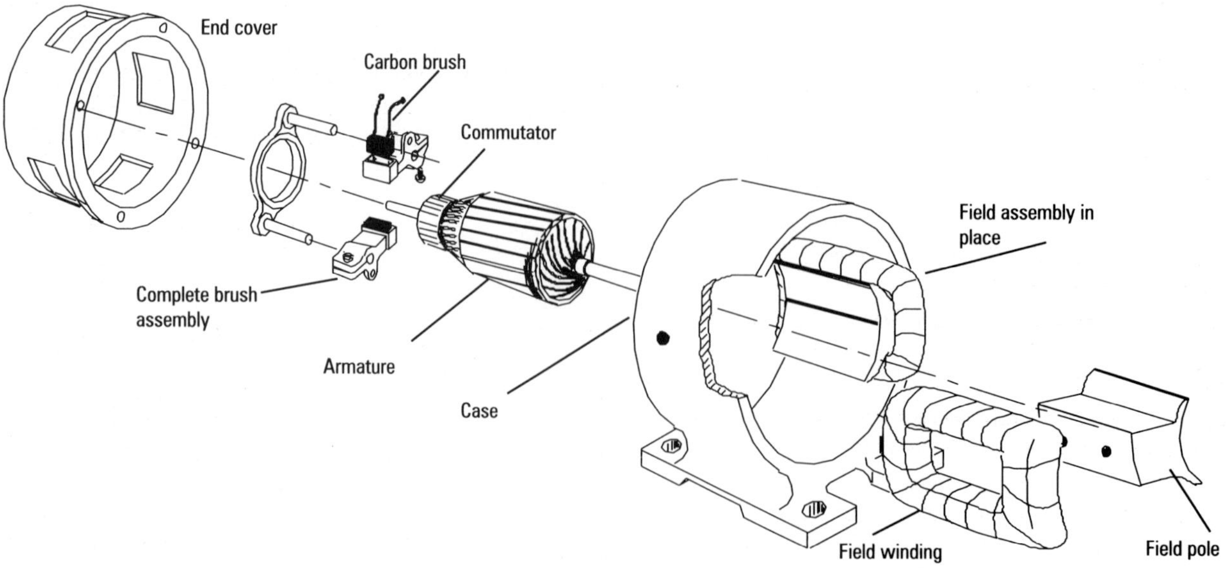

Figure 3.27 Construction of a d.c. generator

Materials used in construction

Case   iron, to complete the magnetic circuit
Armature core   magnetic steel laminations, to reduce
                        hysteresis and eddy currents
Commutator   copper segments
Brushes   carbon composition, negative
                        temperature/resistance coefficient
Field windings   copper wire formed into coils
Field poles   cast iron ground into shape

Construction

The basic construction of d.c. generators is the same as d.c. motors. Figure 3.27 shows an exploded view of a typical two pole generator. The field windings are preformed to the shape of the pole they are to fit. The pole is bolted to the case of the machine so that it forms part of the magnetic circuit. On generators the pole shoe is made of solid iron so that some residual magnetism is held when the machine is switched off. The armature and brush arrangements are the same as for motors.

The theory of the generator is that the armature is rotated by an external prime mover. This may be in the form of an engine or a method of using natural resources such as the wind. As the armature conductors rotate they move through the magnetic field produced by the field windings. This magnetic field will vary in strength depending on where it is supplied from. Separately excited machines, which take their field supply from an external d.c. source, have a strong magnetic field from stationary. Self excited generators have to rely on the residual magnetism left in the poles to get the first e.m.f.s generated.

When the armature conductors move through the magnetic field an e.m.f. is induced into the armature windings

$$E \propto \Phi N$$

,

It can be seen from this that the strength of the magnetic flux has a direct relationship to the generated e.m.f.

The applications of the generators varies with the characteristics for the different types. Some are used far more than others. The series generator, for example, has very few applications and is therefore seldom found in use these days.

Separately excited generators

As the name implies the field windings have an external d.c. power supply connected to them to provide the magnetic flux. This supply is often a battery unit which may have a variable resistor incorporated into the circuit to vary the output. The generated output of the machine is taken directly from across the armature.

The output of this type of generator can be fairly constant but the voltage tends to drop off slightly as the load current is increased. This is due to the effect of the load current and armature resistance causing a voltage drop in the armature.

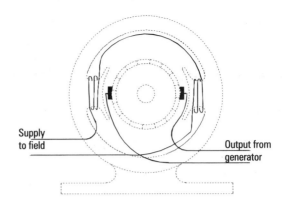

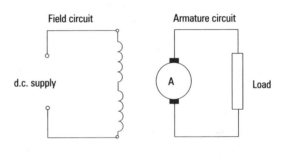

Figure 3.28   *Wiring diagram and circuit diagram of separately excited generator*

The characteristics for this type of machine are shown in Figure 3.29.

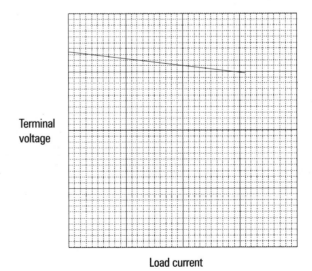

Figure 3.29   *Load characteristics of a separately excited generator.*

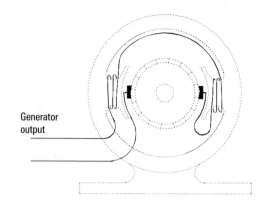

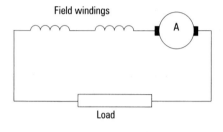

Figure 3.30   *Wiring diagram and circuit diagram of series connected generator*

Self excited generators

Unlike the separately excited generators there is no external supply in this type to power the electromagnets that produce the field. The field windings receive their supply from that produced by the armature. This means that there must be some residual magnetism in the poles for a start so that an initial voltage can be created. Once the armature is turning and current starts to flow in the circuits the magnetic field becomes stronger and the supply voltage increases. If there is no residual magnetism to start with no voltage can be generated. Similarly if the machine is started so that the residual magnetism is weakened instead of strengthened, it will not excite and therefore not generate.

Examples of self excited generators are:
- the series wound generator
- the shunt wound generator
- the compound wound generator

Series wound generators

In a series connected generator the current flowing in the field windings is the same as that in the armature. The field winding is therefore made of a comparatively few turns of very thick wire.

When the current being supplied by the generator is small, the current flowing in the field is small and the magnetic field is weak. This results in a low voltage being generated. As the load current is increased so the field strength is increased and a higher voltage is generated. This can be seen in the characteristics shown in Figure 3.31.

On no load there is a small voltage generated due to the residual magnetism in the poles. The use of this type of generator is very limited but in the past they have been used to boost the voltage to overcome voltage drop in long cable runs.

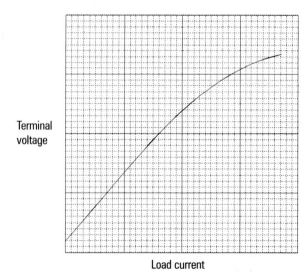

Figure 3.31   *Load characteristics of a series connected generator.*

The shunt wound generator

As can be seen from Figure 3.32 the field and armature windings are connected in parallel (shunt) with each other. This means that the strength of the magnetic field is determined by the resistance of the field windings and the generated voltage. In practice the field winding is made up of coils with a large number of turns with comparatively thin wire. The power absorbed by the field is usually only about 2% of the generator's rated output.

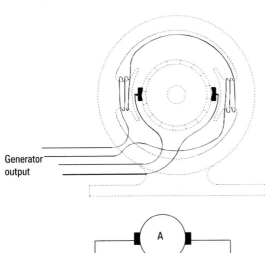

Figure 3.32   *Wiring diagram and circuit diagram of shunt wound generator*

The terminal voltage of this type of generator is fairly constant in a short range of loads. If the load is increased beyond this a voltage drop starts to appear. At first this is due to the resistance of the armature windings but as a drop in terminal voltage also affects the field windings, a reduced magnetic field results. This in turn means that less voltage is generated so an even smaller terminal voltage is produced. As the characteristics show in Figure 3.33 if the load current is significantly increased the terminal voltage continues to drop off.

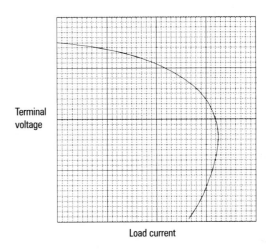

Figure 3.33   *Load characteristics of a shunt wound generator*

The terminal voltage can be controlled to some extent using a field connected regulator. By varying the resistance of the field circuit the current can be controlled and hence the output voltage (Figure 3.34).

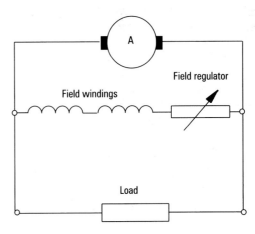

Figure 3.34   *Circuit diagram showing a field regulator in a shunt wound generator circuit.*

Compound wound generator

The compound wound generator, as with the similarly wound motor, has both shunt and series connected windings. As can be seen from Figure 3.35 there are two ways of connecting the shunt windings. A short shunt connected machine takes the shunt winding directly across the armature, whereas a long shunt type takes it across the output terminals. In practice this makes very little difference as the series turns are very few and made of comparatively thick wire, unlike the shunt windings that consist of a large number of turns of comparatively thin wire. In either case the two sets of coils are usually connected so that their ampere-turns assist each other.

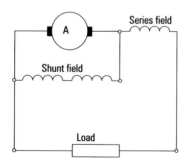

Figure 3.35a   *Circuit diagram of a short shunt compound wound generator*

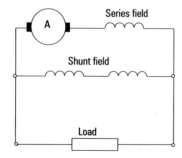

Figure 3.35b   *Circuit diagram of a long shunt compound wound generator*

A generator that is designed and connected so that the currents set up in both the series and shunt windings assist each other is said to be "cumulatively compounded". If the shunt winding is the dominant one and the series has little effect, this is described as being under-compounded as shown in Figure 3.36 (curve A). To level out the curve more series ampere-turns need to be introduced, as curve B in Figure 3.36. In some instances generators are "over-compounded" to meet particular applications. The characteristic for these are shown in Figure 3.36 (curve C).

Generators that are designed and connected so that the field currents oppose each other are called 'differentially compounded' and have characteristics as shown in Figure 3.36 (curve D).

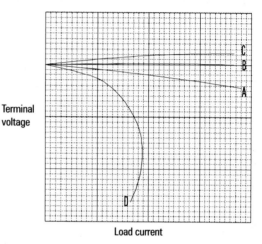

Figure 3.36   *Characteristics of compound generators showing:*
A – under compounded
B – level compounded
C – over compounded
D – differentially compounded

D.C. generator circuits/operating characteristics

The field windings of a separately excited d.c. generator have an external ____ _____ _____ connected to them to provide the _____ _____.

A self excited d.c. generator will not generate an output _____ if there is no_____ magnetism in the field _____.

The current flowing in the field windings of a series wound generator is the same as that flowing in the _____, therefore, the field _____ are made from a _____ gauge wire of comparatively _____ turns.

To reverse the output polarity of a shunt wound generator charge either the _____ or _____ connections but _____ both.

The shunt field winding of a compound wound generator may be connected two different ways:
(a) in parallel with just the _____ for a _____ shunt connection
(b) in parallel with the _____ and the _____ (which are both in _____) For a _____ shunt connection.

Self assessment short answer questions

1. What is the terminal voltage of a d.c. generator which has an armature resistance of 0.25 Ω and a current flowing through it of 5 A when it generates an e.m.f. of 220 V?

2. State
   (a) which type of d.c. motor would be most suitable where large starting torques are required.

(b) the purpose of a field diverter in a series wound motor circuit.

3. Sketch the basic circuit arrangements for each of the following d.c. motors:
   (a) series wound

   (b) shunt wound

   (c) compound wound (long shunt connection)

4. Explain why it is necessary to provide a starter for large d.c. motors.

5. Describe how the direction of rotation can be reversed on the following d.c. motors:
    (a) shunt wound

    (b) compound wound

6. A 250 V d.c. motor has an armature resistance of 0.1 Ω, calculate the starting current if no starting resistor is connected into the armature circuit. Ignore the field resistance.

4

Alternating Current Theory (Single-Phase)

Answer the following questions to remind yourself of what was covered in Chapter 3.

1. A d.c. motor has an armature which has a voltage drop of 6.5 V when connected to a 240 V d.c. supply. Calculate the back e.m.f. of the motor.

2. Sketch a graph showing the speed/torque characteristics of a d.c. shunt motor.

3. State:
   (a) the name of the two electromagnetic devices, used for protection, inside a faceplate starter

   (b) the function of these two devices

4. (a) Which type of mechanical drive is a d.c. shunt-wound motor most suitable for?

   (b) Give two applications of this type of motor.

Part 1

Phasor diagram representation of a.c. quantities

It is often more convenient to represent alternating quantities (for example voltage and current) by a phasor diagram instead of a waveform diagram.

The magnitude (size) of the voltage or current is represented by drawing a straight line to scale or by writing the magnitude of the voltage or current at the side of the line.

The direction of the voltage or current is indicated by the position of the line on the diagram.

The sense of the voltage or current is indicated by an arrowhead on the line.

Example
When an alternating voltage is applied to a circuit it is found that the current produced lags the voltage by 45°. Construct a wave and a phasor diagram for this circuit.

Since the magnitude of the voltage and current are not given it is not necessary to draw the diagrams to scale.

There are two possible answers both equally correct.

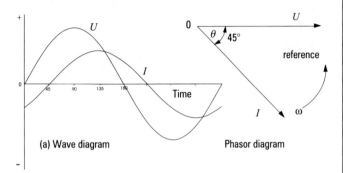

(a) Wave diagram   Phasor diagram

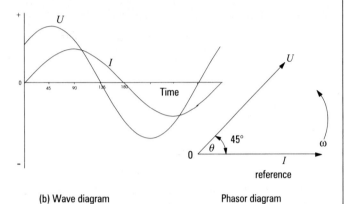

(b) Wave diagram   Phasor diagram

Figure 4.1

The conventional direction of rotation of a phasor is anti-clockwise therefore the current lags behind the voltage by 45° in both cases (a) and (b).

Which do we choose?

Phasor diagram (a) with U as the reference phasor is suitable for a parallel circuit since the supply voltage is the same for each branch in a parallel circuit.

Phasor diagram (b) with I as the reference phasor is suitable for a series circuit since the same current flows in each part of the circuit.

Figure 4.2 shows how "phasor diagrams" can be used to represent the relationship between current and applied voltage in

    (a) a purely resistive circuit
    (b) a purely inductive circuit
    (c) a purely capacitive circuit

Note: The applied voltage is taken as the reference or datum.

(a)
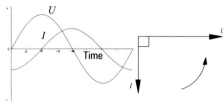

The current is in phase with the voltage.

(b)
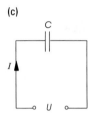 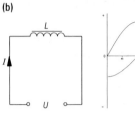

The current lags behind the applied voltage by 90°.

(c)
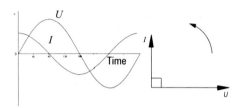

The current leads the applied voltage by 90°.

Figure 4.2

Remember this useful memory aid

C I V I L

In a capacitive circuit the current leads the voltage.

In an inductive circuit the current lags the voltage.

Phasor addition
Alternating quantities cannot be added arithmetically when they are out of phase with each other.

Let's consider the simple R-L series circuit.

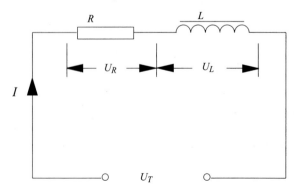

Figure 4.3

The voltage across the resistor U_R cannot be added to the voltage across the inductor U_L to find the total voltage U_T because they are out of phase with each other.

Example

In Figure 4.3 if the voltage across the resistor is 40 V and the voltage across the inductor is 60 V find the value of the supply voltage by drawing a phasor diagram to a scale of 1 cm = 10 V.

Answer

Remember − I is the reference phasor in a series circuit.

Step 1. Draw the current phasor horizontally as "reference" (not to any scale but to a suitable length).

Figure 4.4

Step 2. Draw the phasors for U_R and U_L to scale.

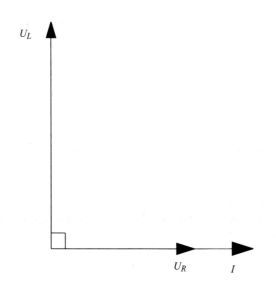

Figure 4.5

Step 3. Construct the phasor parallelogram and hence determine the supply voltage U_T. This is represented by the diagonal line of the parallelogram.

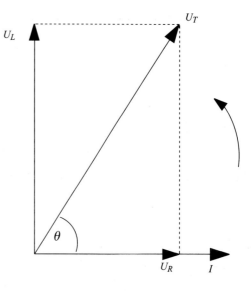

Figure 4.6

Measure the length of the diagonal line and you will find that $U_T = 72V$, which is the "phasor sum" of U_L and U_R.

Note: θ is the number of degrees that I lags behind the supply voltage U_T.

In this example angle $\theta = 56°$

Check this with your protractor.

Applying Pythagoras' Theorem to solve electrical problems

Impedance triangle

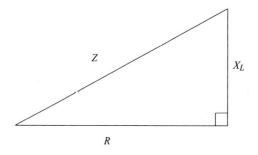

Figure 4.7

Where X_L is inductive reactance in ohms
           R is resistance in ohms
           Z is impedance in ohms

Note: Since X_L and R are always shown at right angles Pythagoras' theorem can be used.

This means that:

$$Z^2 = R^2 + X_L^2 \text{ or}$$

$$Z = \sqrt{R^2 + X_L^2}$$

Example

A resistor of 30 Ω is connected in series with an inductor of reactance 40 Ω. Ignoring any resistance in the inductor, calculate the impedance of the circuit.

$$Z = \sqrt{R^2 + X_L^2}$$

$$= \sqrt{30^2 + 40^2}$$

$$= \sqrt{900 + 1600}$$

$$= \sqrt{2500}$$

$$= 50 \text{ Ω}$$

Try this

Apply Pythagoras' theorem to determine the voltage across the inductance U_L.

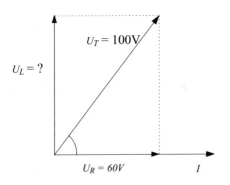

Figure 4.8   Voltage triangle

A.C. circuit calculations

R-L series circuits

Example

A coil has a resistance of 6 Ω and an inductance of 25.5 mH. If the current flowing in the coil is 10 A when connected to a 50 Hz supply determine the supply voltage by drawing a phasor diagram to a scale of 1 cm = 10 V.

(**Note:** a coil is represented by a R-L series circuit.)

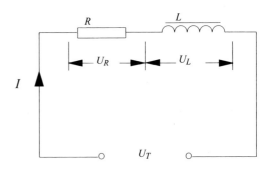

Figure 4.9

Step 1. Calculate the values of U_R and U_L.

$$U_R = I \times R$$
$$= 10 \times 6$$
$$= 60 \text{ V}$$

$$U_L = I \times X_L$$
$$= 10 \times ?$$

We cannot do this until we have found the inductive reactance X_L. To do this we use the formula:

$$X_L = 2\pi f L$$

where f is the frequency of the supply (hertz) and L is the inductance of the coil (henry)

$$= 2 \times 3.142 \times 50 \times 25.5 \times 10^{-3}$$
$$= 8 \text{ Ω}$$

$$U_{\mathrm{L}} \quad = I \times X_L \quad\quad = 10 \times 8 \quad = 80 \text{ V}$$

Step 2. Construct the phasor diagram and measure the diagonal line to find the answer.

Phasor diagram scale 1 cm = 10 V

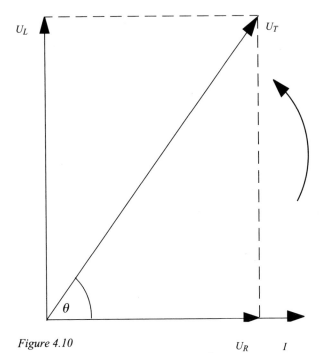

Figure 4.10

Answer = 100V

Impedance – symbol *Z*

It should be seen from the previous example that there are two separate oppositions to the flow of current in an R-L circuit, one is due to the resistance (*R*) and the other is due to the inductive reactance (*X$_L$*). The combination of these oppositions is called the "impedance" of the circuit and it is measured in ohms (Ω).

Remember the impedance triangle.

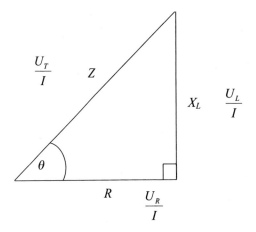

Figure 4.11   Impedance triangle for R-L series circuit

from the diagram:

$$Z \quad = \frac{U_T}{I}, \quad X_L \quad = \frac{U_L}{I} \text{ and } R = \frac{U_R}{I}$$

$$\therefore U_T \quad = \quad IZ, \quad U_L = IX_L \text{ and } U_R = IR$$

$$\therefore I \quad = \frac{U_T}{Z}, \quad I = \frac{U_L}{X_L} \text{ and } I = \frac{U_R}{R}$$

Remember – if the resistance and inductive reactance are known the impedance can be found by the formula

$$Z \quad = \sqrt{R^2 + X_L^2}$$

R-C series circuit

Example
A capacitor of 160 μF is connected in series with a non-inductive resistor of 15 Ω across a 50 Hz supply. If the current drawn is 10 A
(a)   calculate
    (i) the capacitive reactance
    (ii) the voltage across each component
(b)   find by means of a phasor diagram the supply voltage.

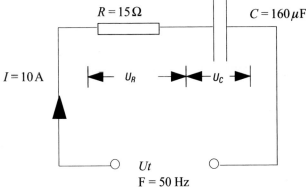

Figure 4.12

(a) (i) $X_C = \dfrac{10^6}{2\pi fC} = \dfrac{10^6}{2 \times 3.142 \times 50 \times 160} = 19.89\ \Omega$

Say 20 Ω

(ii) $U_R = I \times R = 10 \times 15 = 150$ V
$U_C = I \times X_C = 10 \times 20 = 200$ V

(b)

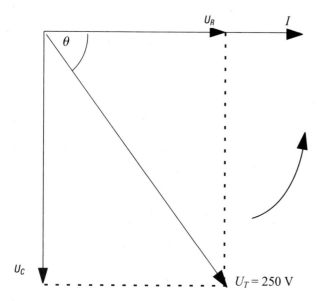

Figure 4.13

Phasor diagram scale 1 cm = 30 V

Typical waveform diagrams for:
R-L series circuit (Figure 4.14)

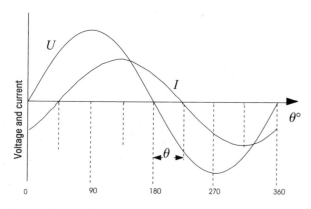

Figure 4.14   Note: the current lags behind the voltage by the "phase angle" θ.

R-C series circuit (Figure 4.15)

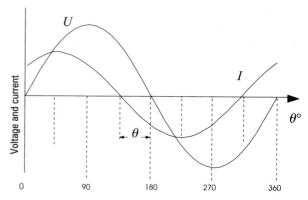

Figure 4.15   Note: the current leads the voltage by the "phase angle" θ.

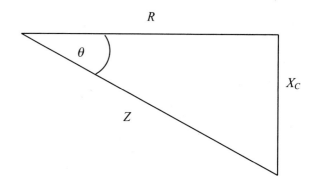

Figure 4.16   Impedance triangle for R-C series circuit.

R-L-C series circuits

In the R-L-C series circuit there are three oppositions to the flow of current; resistance, inductive reactance and capacitive reactance.

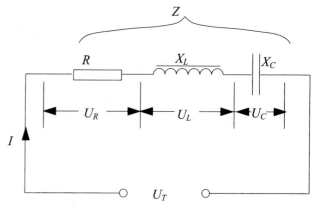

Figure 4.17

If the inductive reactance is the same as capacitive reactance, the impedance of the circuit is the same as the resistance of the circuit.

i.e. if $X_L = X_C$ then $Z = R$

The impedance of the circuit can be found by the formula

$$Z = \sqrt{R^2 + (X_L - X_C)^2}$$

Consider the circuit below

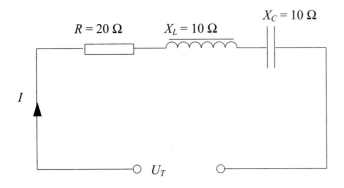

Figure 4.18

$$Z = \sqrt{R^2 + (X_L - X_C)^2}$$
$$= \sqrt{20^2 + (10 - 10)^2}$$
$$= \sqrt{20^2}$$
$$= \sqrt{400}$$
$$= 20 \ \Omega$$

We have proved that $Z = R$ when $X_L = X_C$.

Note: Power factor $= \dfrac{\text{resistance}}{\text{impedance}} = \dfrac{R}{Z}$

and when $Z = R$ the power factor of the circuit is unity.
Therefore the power factor of the circuit in Figure 4.18

$$= \frac{R}{Z} = \frac{20}{20} = 1 \quad \text{(Power factor is covered in more}$$

detail later in this chapter.)

Phasor diagram for R-L-C series circuit (Assuming that U_L is greater than U_C)

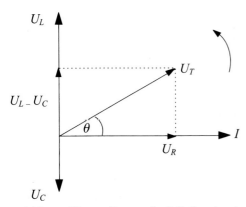

Figure 4.19   Phasor diagram for R-L-C series circuit

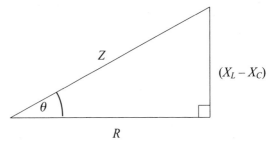

Figure 4.20   Associated impedance triangle

 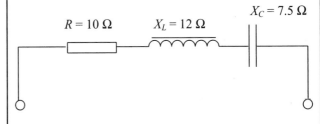

Periodic time and frequency

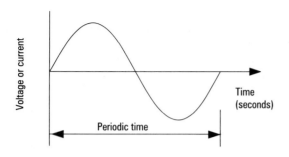

Figure 4.22   Full cycle of an alternating voltage or current

The time it takes to complete one full cycle of the a.c. supply is known as the periodic time (*T*).

The number of full cycles completed in one second is known as the frequency (*f*).

The supply frequency can be calculated using the formula:

$$f = \frac{1}{T}$$

where *f* is frequency in Hertz (Hz) and *T* is periodic time in seconds (s)

Example
What is the periodic time of a 50 Hz a.c. supply?

$$T = \frac{1}{f} = \frac{1}{50} = 0.02 \text{ s or } 20 \text{ ms}$$

Try this
Determine the frequency of the waveform shown.

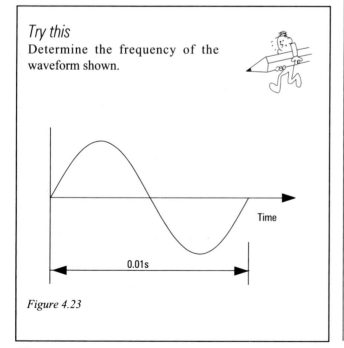

Figure 4.23

Phasors of a.c. quantities/a.c. series circuits

I is used as the reference on a phasor diagram for a _____ circuit since the _____ current flows in each part of the circuit.

U is used as the reference on a phasor diagram for a _____ circuit since the _____ _____ is the same for each branch of the circuit.

In a capacitive circuit the current _____ the voltage, and in an inductive circuit the current _____ the _____.

The impedance of a circuit containing resistance and inductive _____ can be found by using the formula:

_____

The impedance triangle for an R-C series circuit has:

_____ on the horizontal side

_____ on the vertical side, and

_____ on the hypotenuse.

The _____ of a series R-L-C circuit can be found by the formul

a $Z = \sqrt{R^2 + (X_C - X_L)^2}$

when the _____ is greater than the _____.

The time taken to complete one full cycle of the _____ supply is known as the _____ _____.

Part 2

Applying trigonometrical ratios to solve electrical problems

Remember – SOH CAH TOA

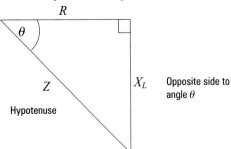

Figure 4.24

$$\sin \theta = \frac{\text{opp}}{\text{hyp}} \qquad \textbf{(SOH)}$$

$$\cos \theta = \frac{\text{adj}}{\text{hyp}} \qquad \textbf{(CAH)}$$

$$\tan \theta = \frac{\text{opp}}{\text{adj}} \qquad \textbf{(TOA)}$$

Example
In Figure 4.24 if
    the angle $\theta = 45°$ and
    the inductive reactance $X_L = 100\ \Omega$
determine the values of R and Z (in ohms).

Since $\sin \theta \quad = \dfrac{\text{opp}}{\text{hyp}} = \dfrac{X_L}{Z}$

        $\therefore Z \quad = \dfrac{X_L}{\sin \theta}$

$(\sin 45° = 0.707)$

            $= \dfrac{100}{0.707}$

            $= 141.4\ \Omega$

Since $\tan \theta \quad = \dfrac{\text{opp}}{\text{adj}} = \dfrac{X_L}{R}$

        $R \quad = \dfrac{X_L}{\tan \theta}$

$(\tan 45° = 1)$

        $R \quad = \dfrac{100}{1} = 100\ \Omega$

Notes: (1) Values of sin, cos and tan can be obtained from any scientific calculator. Examples: sin 36.87° = 0.6, cos 36.87° = 0.8 and tan 36.87° = 0.75.

(2) If you know the values of sin, cos and tan you can find the angle from your calculator. Examples: $\sin^{-1} 0.6 = 36.87°$, $\cos^{-1} 0.8 = 36.87°$ and $\tan^{-1} 0.75 = 36.87°$.

Try this

1.

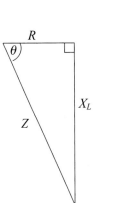

Figure 4.25

If the resistance $R = 15\ \Omega$ and the impedance $Z = 30\ \Omega$ determine
(a) the power factor
(b) the angle θ
Note: power factor = the cosine of the angle θ

2.

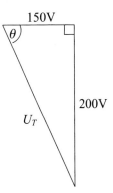

Figure 4.26

(a) Using Pythagoras' theorem determine the value of U_T.
(b) Using trigonometry determine the angle θ.

Power factor

The product of voltage and current gives the "true power" of a d.c. circuit (i.e. $P = U \times I$), and this is the amount of power actually consumed by the circuit. In an a.c. circuit the product of voltage and current ($U \times I$) is the "apparent power" and this is only equal to the "true power" if the voltage and current are in phase with each other, as in a purely resistive a.c. circuit (Figure 4.27).

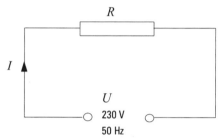

Figure 4.27   Circuit diagram

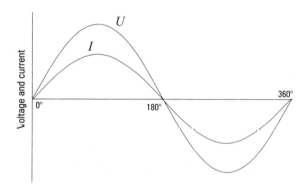

Figure 4.28   Waveform diagram showing U and I in phase with each other.

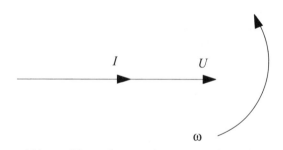

Figure 4.29   Phasor diagram showing U and I in phase.

The majority of a.c. circuits, such as motor circuits, contain a combination of inductance and resistance so that the voltage and current are out of phase with each other (Figure 4.30) and the product gives apparent power.

(a) Single-phase a.c. circuit

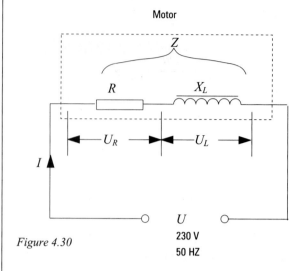

Figure 4.30

Motor circuits are usually represented as resistance (R) and inductive reactance (X_L) in series. The result of this combination is the impedance (Z).

The current (I) flows through both R and X_L as they are in series and therefore is common to each. It is also in phase with the voltage across the resistance (U_R) whereas the voltage across the inductive reactance (U_L) is out of phase by 90°.

(b) Resultant waveform diagram

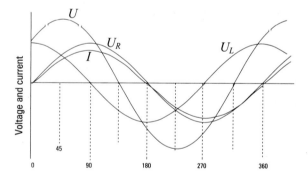

Figure 4.31

(c) Phasor diagram

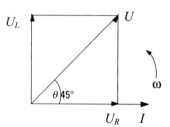

Figure 4.32

The phasor diagram shows the phase relationships more clearly.
Note: The current lags behind the voltage (U).

The ratio of true power to apparent power is known as the **POWER FACTOR**. This can be found by constructing a "power triangle" and applying trigonometry.

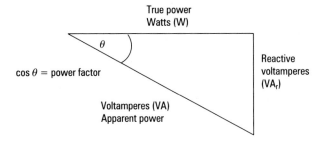

Figure 4.33   *Power triangle – when power factor is lagging*

The inductive part of the circuit (Figure 4.30) is called the "Wattless Component" (VA $_r$), since it consumes no power. It only provides a magnetic field. The resistive part is called the "Wattful Component" or "True Power" (W) since all the power is dissipated in the circuit resistance. The combination of the two is known as the "Apparent Power" (VA).

By trigonometry:

$$\cos \varnothing = \frac{\text{adjacent}}{\text{hypotenuse}}$$

$$\cos \theta = \frac{\text{true power}}{\text{apparent power}} = \frac{W}{VA}$$

∴ Power factor = Cosine of the angle (θ)

Note:
The power triangle is usually shown in terms of kVA, kW and kVA $_r$ (Figure 4.34).

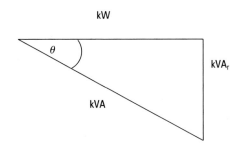

Figure 4.34

Now power factor = $\cos \theta = \dfrac{kW}{kVA}$

Power factor can be one of three conditions, namely:
• unity power factor
• lagging power factor
• leading power factor

When the true power = apparent power, the power factor is unity (1), and the wattless power (kVA $_r$) is zero.

Power factors in **inductive** circuits are termed **"lagging"** as the current lags the voltage

Power factors in **capacitive** circuits are termed **"leading"** as the current now leads the voltage.

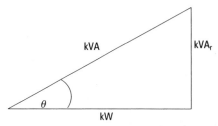

Figure 4.35   *Power triangle – when power factor is leading*

Remember

Motor circuits, transformer circuits and so on comprise resistance and inductance therefore they have lagging power factors.

Power factor is the ratio of true power to apparent power.

Applying trigonometry:

Power factor = $\cos \theta = \dfrac{W}{VA}$ or $\dfrac{kW}{kVA}$

Example
A 30 Ω resistor and an inductive reactance of 40 Ω are connected in series to a 230 V supply.

Calculate:
(a)   the apparent power
(b)   the power factor
(c)   the true power
(d)   the reactive power

(a) $Z = \sqrt{R^2 + X_L^2}$

$\quad\quad = \sqrt{30^2 + 40^2}$

$\quad\quad = 50 \ \Omega$

$\quad I = \dfrac{U}{Z} = \dfrac{230}{50} = 4.6 \ \text{A}$

apparent power = $UI = 230 \times 4.6 = 1058$ VA

(b)   Power factor = $\dfrac{R}{Z} = \dfrac{30}{50} = 0.6$ lagging

(c)   True power = apparent power × power factor
$\quad\quad\quad\quad\quad = 1058 \times 0.6 = 634.8$ W

(The same as $P = UI \cos \theta = 634.8$ W)

(d) Reactive power $= \sqrt{(VA)^2 - W^2}$

$$= \sqrt{1058^2 - 634.8^2}$$

$$= 846.4 \text{ VA}_r$$

Try this

A 15 Ω resistor and a capacitive reactance of 30 Ω are connected in series to a 230 V, 50 Hz supply.

Calculate the apparent power, the power factor, the true power and the reactive power.

Power factor measurement

Power factor can be measured by

(a) using a power factor meter.

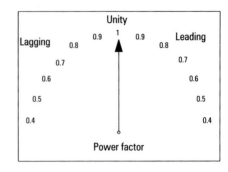

Figure 4.36   Scale of a power factor meter

Power factor meters are connected into circuits in a similar way to wattmeters.

(b) using a wattmeter, voltmeter and ammeter connected as shown in Figure 4.37.

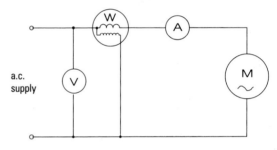

Figure 4.37

The wattmeter measures the "true power", and the voltmeter and ammeter arrangement measures the "apparent power".

Try this

If the wattmeter reads 5 kW, the voltmeter reads 230 V and ammeter reads 32 A calculate the power factor of the motor.

Power factor

Power factor can be found from (a) the _____ triangle and (b) the _____ triangle by applying trigonometry.

(a)

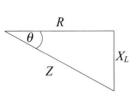

(b)

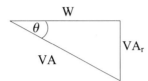

Figure 4.38

Here P.F. = cos θ = _____

Figure 4.39

Here P.F. = cos θ = _____

The formula $P = U \times I$ gives the:

(i) _____ of a d.c. circuit
(ii) the _____ of a _____
resistive _____ circuit

The ratio of true power to _____ power is known as the _____ _____ .

Power factor in _____ circuits are termed "lagging" as the _____ lags the _____ .

Power factors in _____ circuits are termed "leading" as the _____ now leads the _____ .

The power factor of a circuit is unity when:

$X_L =$ _____
and $Z =$ _____

Power factor can be measured using the following meters

1. _____

2. _____

3. _____

Part 3

Improving the power factor of an a.c. motor

Power factor can be improved by connecting a capacitor across the supply in parallel with the motor (Figure 4.40).

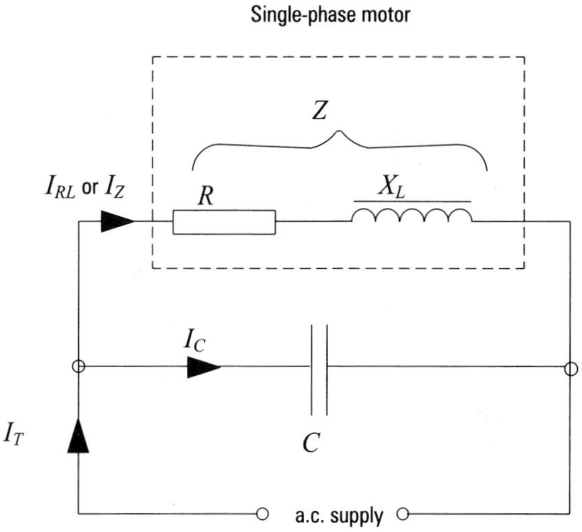

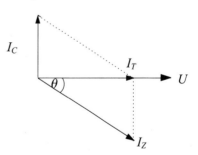

Figure 4.40

C is a power factor correction capacitor
I_{RL} or I_Z = the current through the motor impedance
I_C = capacitor current
I_T = total current

The capacitor effectively brings the total current nearer into phase with the supply voltage. The leading power factor of the capacitor improves the lagging power factor of the motor inductance.

Figure 4.41

In Figure 4.41 I_Z is the result of the resistance and the inductance.

When I_C is added to the circuit it cancels out the reactive effect of the motor.

In the example shown the power factor would be unity as the total current is in phase with the voltage.

In practice unity power factor is seldom achieved and a power factor of 0.8 to 0.9 is considered to be satisfactory.

65

If the power factor is not corrected to unity the total current will not be in phase with the voltage. This means there is a second phase angle introduced.

If the capacitor in the circuit shown in Figure 4.40 does not correct the power factor to unity the phasor diagram will be as in Figure 4.42.

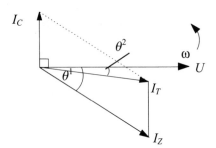

Figure 4.42

As the branch of the circuit containing the capacitor is purely capacitive I_C leads the voltage by 90°.

θ^1 is the angle that the current I_Z lags the voltage by.

θ^2 is the angle that the resultant current I_T lags the voltage by.

If $\theta^1 = 35°$ then $\cos \theta^1 = 0.82$
and $\theta^2 = 13°$ then $\cos \theta^2 = 0.97$

The power factor in this case is improved from 0.82 to 0.97.

Why improve the power factor?

Electrical energy supplied at a low power factor is costly to the supply authority since the larger currents taken from the supply require heavier cables and switchgear than that which is really necessary.

The industrial consumer also pays more than the true power used because their kilovolt ammeter (kVA meter) records the instantaneous values of the product of voltage and current.

Therefore it is in the interests of both supply authority and consumer to keep the power factor as near as possible to unity.

Remember
Capacitors are connected across the supply terminals of a motor to improve the power factor.

The leading power factor of the capacitor counteracts the lagging power factor of the motor inductance and keeps the circuit currents down to acceptable levels.

If the power factor is poor then the supply currents are larger and the cables and switchgear need to be heavier.

Power factor correction

Correction to unity
A 230 V, 50 Hz induction motor takes a current of 50 A and its power factor at this load is 0.8 lagging. It is required to find the value of the capacitor to correct the power factor to unity.

First it will help to draw a simple circuit and to put all the known values on the diagram.

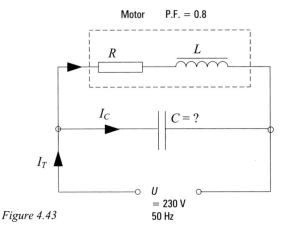

Figure 4.43

As with most of these problems, there are different methods we can use to solve them. Two methods are given here.

Method 1
Graphically, by constructing a phasor diagram to scale.

Construction of the phasor diagram.
(a) Draw the horizontal reference U (not to scale).
(b) Find the angle θ.
     $0.8 \cos^{-1} = 36.87°$ on your calculator
     (on some calculators you will need to press 0.8 INV cos)
(c) I_{RL} can now be drawn to scale at the angle θ using a ruler and protractor.
(d) Draw
     (i)   a vertical line to meet the reference and label it I_L.
     (ii) another vertical line at 90° to U the same length as (i) and label it I_C.
(e) The parallelogram may now be completed.

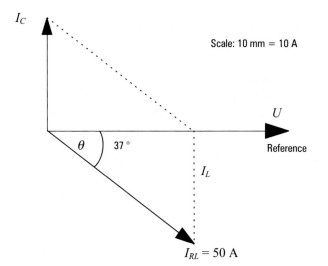

Scale: 10 mm = 10 A

θ 37°

U

Reference

I_L

I_{RL} = 50 A

Figure 4.44

Measure the length of the vertical line which represents the leading current the capacitor must produce.

$$I_C = 30 \text{ Amps}$$

Next find the capacitive reactance (X_C) of the capacitor.

$$X_C = \frac{U}{I_C} = \frac{230}{30} = 7.66\ \Omega$$

Since $$X_C = \frac{10^6}{2\pi f C}$$

the value of the capacitor (in microfarads) can now be found by transposing the formula for C.
(Note: f is the supply frequency)

$$C = \frac{10^6}{2\pi f X_C}$$

$$= \frac{10^6}{2 \times 3.142 \times 50 \times 7.66}$$

$$= 415.5\ \mu\text{F}$$

Method 2
By applying Trigonometry
First sketch a phasor diagram (not to scale).

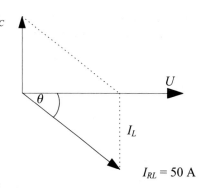

I_{RL} = 50 A

Figure 4.45

Next find the angle θ
Since power factor = $\cos \theta = 0.8$
$\theta = 0.8 \cos^{-1} = 36.87°$

Now find the sine of the angle.
$\sin = \sin 36.87° = 0.6$
(36.87 SIN = 0.6 on your calculator)

By trigonometry

$$\sin = \frac{\text{opposite}}{\text{hypotenuse}}$$

$$\therefore \sin \theta = \frac{I_C}{I_{RL}}$$

Transposing $I_C = I_{RL} \sin \theta$

$$I_C = 50 \times 0.6$$
$$= 30 \text{ amps}$$

The calculation can now be completed as in Method 1.

Try this
The phasor diagram below represents to scale the current taken by a 230 V, 50 Hz motor. A capacitor is connected across the terminals of the motor to raise the power factor to unity.
(a) Use the diagram to find the current taken by the capacitor.
(b) Calculate the capacitance of the capacitor.

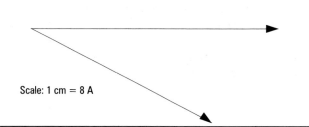

Scale: 1 cm = 8 A

Figure 4.46

A.C. power

To calculate the power in a single-phase a.c. circuit which contains resistance and inductance we must

remember: $\cos\theta = \dfrac{\text{true power}}{\text{apparent power}}$

which when transposed:

true power $= \text{apparent power} \times \cos\theta$

and since

apparent power $= \text{VA (which is simply } U \times I)$

true power $= U \times I \times \cos\theta \text{ (watts)}$

It can be seen that the power in a single-phase a.c. circuit can be calculated by $P = UI\cos\theta$, where θ is the angle by which the current and voltage are out of phase.

We can also calculate the circuit current by transposing $P = U I \cos\theta$ for I as shown below:

$$I = \frac{P}{U \times \cos\theta}$$

Try this

Calculate the current drawn by a 230 V, 3 kW single-phase load with a power factor of 0.8 lagging.

Example

At a full load power of 500 W an induction motor running on a 230 V 50 Hz supply has a lagging power factor of 0.7.

Calculate the value of capacitor required to raise the overall power factor to unity.

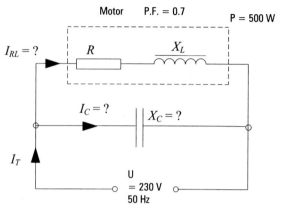

Figure 4.47   Circuit diagram

Since

$P = UI\cos\theta$

$I = \dfrac{P}{U \times \cos\theta}$

$= \dfrac{500}{230 \times 0.7}$

$= \dfrac{500}{161}$

$= 3.1 \text{ A}$

$\therefore I_{RL} = 3.1 \text{ A}$

$0.7 \cos^{-1} = 45.57°$

$\therefore \theta = 45.57°$

$\sin\theta = \sin 45.57° = 0.714$

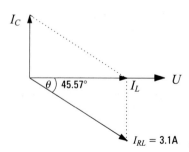

Figure 4.48   Phasor diagram

Since we now have I_{RL} and the angle we can calculate I_C. As this uses the Sine of the angle this should first be found.

$$\sin \theta = \frac{I_C}{I_{RL}}$$

$$\therefore \quad I_C = I_{RL} \sin \theta$$

$$= 3.1 \times 0.714$$

$$= 2.2 \text{ A}$$

Now we have I_C we can find X_C from:

$$X_C = \frac{U}{I_C} = \frac{230}{2.2} = 104.5 \, \Omega$$

Since $\quad X_C = \frac{10^6}{2\pi f C}$

$$C = \frac{10^6}{2\pi f X_C}$$

$$= \frac{10^6}{2 \times 3.142 \times 50 \times 104.5}$$

$$= \frac{10^6}{32833.9} = 30.456 \, \mu\text{F}$$

Points to remember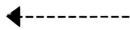

Power factor improvement/correction

Power factor can be improved by connecting a _____ across the _____ in parallel with an inductive load.

When power factor is poor the supply current is _____ and cables, and switchgear need to be _____.

The capacitance of a capacitor (in microfarads) can be found using the formula:

_____

The true power of a single-phase a.c. circuit containing resistance and inductance can be found using the formula:

$$P =$$

and the circuit current can be found by transposing the formula, therefore

$$I =$$

1. Sketch an appropriate phasor diagram (not to scale) for EACH of the following a.c. circuits.

   (a)

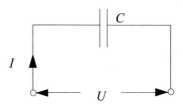

Figure 4.49(a)

   (b)

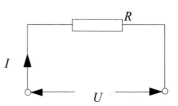

Figure 4.49(b)

   (c)

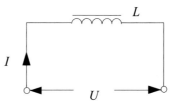

Figure 4.49(c)

2. Draw a phasor diagram to scale, for the circuit shown below, and determine the value of the supply current from the phasor diagram.

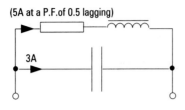

Figure 4.50

3. Calculate
   (a) the impedance and
   (b) the current for the circuit shown below.

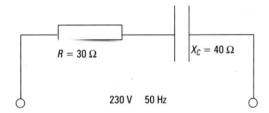

Figure 4.51

4. For the circuit shown below, calculate the
   (a) power factor
   (b) power dissipated

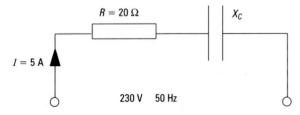

Figure 4.52

5. (a) Give TWO reasons for improving the power factor of a motor circuit installation.
   (b) Show by means of a sketch, how the power factor of a single-phase motor can be improved.

6. A single-phase a.c. load draws a current of 6 A from a 230 V, 50 Hz supply. If the power factor of the load is 0.85 lagging, calculate the power input.

7. The values indicated by the instruments connected into an a.c. circuit on load were:
   voltmeter 235 V
   ammeter 7.2 A
   wattmeter 1500 W
   (a) Calculate the power factor of the circuit.
   (b) Assuming that the load is "inductive" determine the phase angle by which the current "lags" behind the supply voltage.

8. Draw the waveform diagram for an a.c. circuit containing resistive and inductive components connected in series, indicating voltage, current and angle of phase difference.

5

Alternating Current Theory
(Three-Phase)

Answer the following questions to remind yourself of what was covered in Chapter 4.

1. Draw an impedance triangle for an a.c. circuit having resistance and inductance connected in series, and identify each side of the triangle.

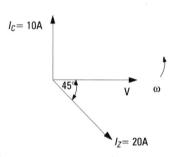

Figure 5.1

2. By means of a waveform diagram, show the relationship between voltage and current in a purely capacitive a.c. circuit.

3. A series circuit consists of a pure resistor of 20 Ω and an inductor of negligible resistance. When connected to a 230V, 50 Hz supply, the circuit current is 5 A. Calculate the voltage across the
    (a) resistor
    (b) inductor

4. The phasor diagram shown Figure 5.1 is for an R-L series circuit in parallel with a capacitor. Redraw the phasor diagram to a suitable scale and from it determine the total current taken from the supply.

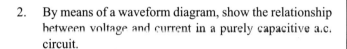

On completion of this chapter you should be able to:

◆ describe the generation, transmission and distribution of electrical energy by a three-phase system

◆ identify and calculate voltages and currents in both star and delta three-phase circuits

◆ draw circuit, waveform and phasor diagrams for star connected three-phase circuits under balanced conditions

◆ perform simple calculations for three-phase a.c. circuits/loads

◆ explain the reasons for balancing single-phase loads on three-phase systems

◆ explain with the aid of diagrams the production of a rotating magnetic field from a three-phase supply

◆ calculate percentage voltage drops throughout an installation

◆ describe with the aid of diagrams methods used to produce d.c. supplies derived from a.c. sources

Part 1

Generation, transmission and distribution of electrical energy

The constant supply

Whenever we go to switch on electrical equipment we expect the supply to be there. This means that electrical generators must be working 24 hours a day, every day of the year. Mains electricity is not stored, it is generated all the time. To make sure there is enough power available for any anticipated load, generators over the whole country have to be running regardless of whether the supply is used.

Not only have the generators got to be running but there must also be an electrical connection between the generator and to where the electricity is to be used. The cables that make this connection often have to be hundreds of miles long. To ensure the most efficient use of these cables the voltage is varied several times before the supply reaches the consumer.

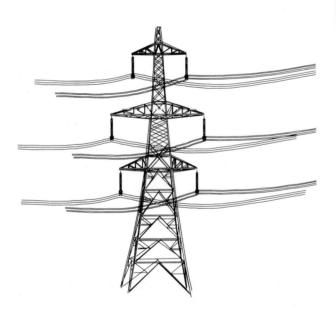

Figure 5.2

Three-phase generation

The three-phase generator (also known as a three-phase alternator) has three similar stator coils (R, Y, B) placed as shown in Figure 5.3 and when the rotor is rotated an e.m.f. will be generated (induced) in each coil. This generated e.m.f. will have the same frequency and maximum (peak) value, however, owing to the relative displacements of the coils the maximum values of the three e.m.f.'s will occur at different instants. Since the three coils are spaced 120° apart, then the waveforms of the e.m.f.'s induced in the coils will be displaced by 120° (see Figure 5.4)

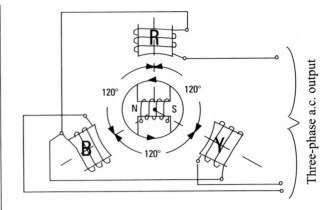

Figure 5.3   *Simple three-phase (star connected) a.c. generator*

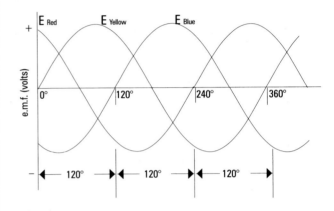

Figure 5.4   *Generation of three-phase e.m.f.*

The output waveform of the generator (which is simply 3 sine waves displaced 120° out of phase with each other) may be represented by a rotating phasor as shown in Figure 5.5.

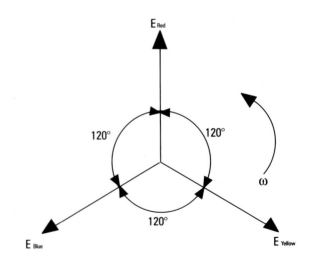

Figure 5.5   *Three-phase phasor diagram*

Generation to transmission

The output voltage of a generator set in a power station will not be much greater than 25 000 volts (25 kV) a.c., and on older generators it may be considerably less.

For transmission purposes the generated output must be stepped up to a maximum of 400 000 volts (400 kV) or 275 000 volts (275 kV) in some situations.

Remember

Electricity is generated as alternating current.

This means that the voltage can be transformed up or down for transmission and distribution to where the load is situated.

The grid system.

The system of high voltage transmission is known as the Grid System. (Figure 5.6). This consists of a lattice work of cables which connect power stations and large load areas together.

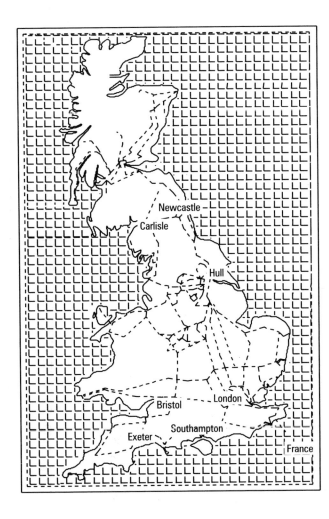

Figure 5.6   The grid system in the U.K.
For clarity only the main lines have been shown.

Power stations are not always situated where the highest loads are. This means that there must be lines from the power stations to all of the load points. There must also be some interconnections between power stations so that they can cover for each other.

Throughout a country not all peak loads appear at the same time and a power station in one area may be called on to supply power to another. From time to time power stations have to be closed down for maintenance and repair and at these times other power stations must cover the total demand.

Over the last 30 years the trend has been to build bigger and more efficient power stations and close the smaller uneconomic ones down.

In addition to the grid system in the U.K. there are cross channel links with the French supply system. These are underwater cables going from the Kent coast to North East France. The time difference in the two countries is one of the factors that makes these links practicable. The peak load times vary thus allowing the U.K. to import power from France when the demand requires it. It also allows the opposite to take place when the French demand is high. For practical reasons these cross channel cables are supplied with d.c. and this is converted to a.c. at each end.

Why high voltage transmission?

Cables transmitting power over long distances need to be kept as small as possible. Conductors with large cross-sectional areas would involve far bigger pylons to take the extra weight and probably more of them. So conductors are kept as small as reasonably possible by using high voltages.

Let's look at an example of how this works.

The average household probably has a load of about 15 kW.

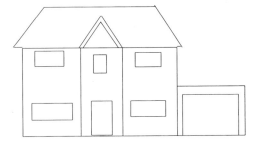

Figure 5.7

A town with 10 000 houses would have a demand of

$$10\ 000 \times 15 = 150\ 000 \text{ kW}.$$

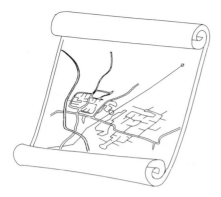

Figure 5.8   Town map

We can assume that this power would be split over three phases, so each phase would carry approximately 50 000 kW.

If this was supplied by cables at 230 V the conductor would need to be capable of carrying

$$\frac{50\,000\,kW}{230\,V} = \frac{50\,000\,000\,W}{230\,V} = 217\,391\,A$$

a very large cable would be required!

However if this same load is supplied by cables at 400 kV the conductors will only need to be capable of carrying

$$\frac{50\,000\,kW}{400\,kV} = \frac{50\,000\,000\,W}{400\,000\,V} = 125\,A$$

A single 400 mm^2 overhead transmission cable which could be used on 400 kV supplies has a current rating of 650 A. These cables are usually bunched in twos or fours on the pylons for more efficient transmission of power.

Loads on a.c. supplies

To simplify the load calculation watts and kW have been used, however, in practice on a.c. supplies the loads would be measured in kVA and not kW. When using d.c. supplies watts can be calculated from the volts multiplied by the amperes. When alternating currents are used it is only on purely resistive loads that this is true. All equipment that uses coils of wire associated with an iron core has what is known as reactance and this results in the voltage and current being out of phase with each other.

If a motor circuit was connected with a voltmeter, ammeter and wattmeter the readings on the voltmeter multiplied by those on the ammeter would not equal that of the wattmeter. In fact they would be greater. The factor that makes up the difference can be calculated from

$$\frac{P}{UA}$$

and this is known as the power factor.

The result of this is that a.c. "power" may apparently be measured in two quantities:

- kW when load is purely resistive
- kVA when the load is not purely resistive

In fact there is a third measurement known as kVAr which relates to the reactive part of a load. These three factors can be represented by a scaled triangle as shown in Figure 5.9.

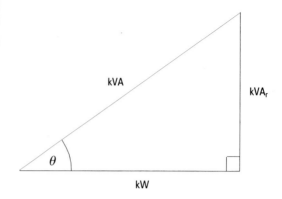

*Figure 5.9   Power triangle (3-phase load)
The cosine of the angle θ represents the power factor (i.e. cos θ = p.f.)*

KW is the base
kVA$_r$ is usually vertical
kVA is the line drawn between the two and is always the highest value.

kVA can be determined by drawing the triangle to scale or by

$$kVA^2 = kW^2 + kVA_r^2$$

or $$kVA = \sqrt{kW^2 + kVA_r^2}$$

Remember
The power triangle for a three-phase load is the same as that for a single-phase load.

Example
A small industrial premises has a three-phase load of 60 kW at 80 kVA$_r$ lagging. Calculate
(a)   the kVA
(b)   the power factor

Answer
(a) $kVA = \sqrt{kW^2 + kVA_r^2}$

$$= \sqrt{60^2 + 80^2}$$

$$= 100\,kVA$$

(b)   Power factor $= \cos \theta = \dfrac{kW}{kVA} = \dfrac{60}{100} = 0.6$ lagging

Try this

A factory has a three-phase load of 50 kVA, 30 kVAr. Determine

(a) the kW

(b) the power factor

Try this

Determine the kVA$_r$ if a large motor has a resistive load of 15 kW and an inductive load of 20 kVA.

Points to remember ◀ ─ ─ ─ ─ ─ ─ ─ ─ ─ ─ ─ ─ ─

Generation, transmission/loads on a.c. supplies

The output of a three-phase generator (also known as a _____ _____ _____) is three similar _____ waveforms displaced by _____ out of _____ with each other.

The output voltage of a generator is approximately _____.

Standard transmission voltages are_____ and _____.

Very_____ transmission voltages are used to keep _____ sizes as small as reasonably possible.

The ratio $\dfrac{kW}{kVA}$ is known as the_____ _____.

The kVA of a three-phase load can be determined by drawing a _____ to _____ or by using the formula:

_____.

Part 2

Distribution

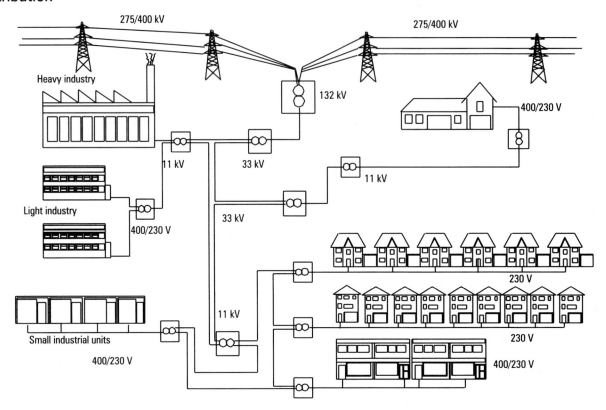

Figure 5.10   *Distribution system showing the voltages used*

Although the systems of distribution we have can sometimes appear to be very complex it is far better than was available in the past.

To give an example in London in 1919 there were eighty separate supply undertakings with seventy different generating stations, fifty different supply systems operating at twenty four different voltages and ten different frequencies. Since those times voltages and frequencies have been standardised.

The electricity supply industry now has 12 regional companies in England and Wales who are also distribution companies responsible for the supply of electricity to the consumers in their area. Each of these purchases their electricity from a generating company. The power is distributed by the National Grid Company which is jointly owned by the 12 Regional Electricity Companies. Scotland and Northern Ireland have their own electricity supply systems.

The Regional Electricity Companies have a legal responsibility to keep the supply within certain limits. Following voltage standardisation in Europe these are that for voltage a nominal supply of 400/230 V, + 10%, -6%; for frequency it must not be more or less than 1% of 50 Hz.

The distribution can be split into three main sections
- industrial
- commercial and domestic
- rural

The reason for splitting these is the different voltages they require and the remoteness of rural supplies. Wherever possible distribution cables are laid underground. Apart from rural areas almost all 11 kV and 400/230 V cables are buried underground and a large number of the higher voltage cables are now also buried.

Figure 5.11 shows the different types of supply which can be obtained from the secondary side of a delta/star transformer.

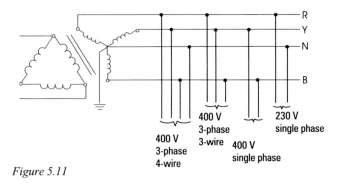

Figure 5.11

Typical uses for each type of supply:

400 V 3-phase 4 wire	– industrial, agricultural and commercial premises
400 V 3-phase 3 wire	– motor circuits
400 V single-phase	– welding plant
230 V single-phase	– small commercial and domestic premises

Industry, if it is large enough, will take a supply at 132 kV. Or in some cases 33 kV. Often if there is an industrial estate of smaller units the estate will have a substation supplied with 132 kV or 33 kV. The transformer will then step-down the voltage to 11 kV for further distribution. In the case of very small units the substation may go directly down to 400/230V without going through another transformer.

Commercial/Domestic supplies are usually at 400/230 V. Large commercial premises will have their own substation transformer fed at 11 kV, which will step down the voltage to 400/230 V for internal distribution.

The 11kV input to the transformer will be connected in "delta" whereas the 400/230V output will be a "star" arrangement. To supply the delta connected windings a three-phase three-wire system is used, with no neutral conductor. The star connected output uses a three-phase four wire connection with the centre point of the star being neutral and connected to earth.

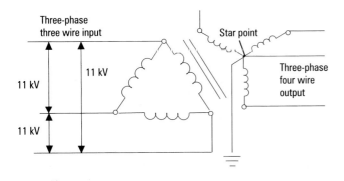

Figure 5.12   11 kV input to the transformer

In delta the voltage across each of the lines is the same as the transformer winding, whereas a transformer winding connected in star will give us a voltage between phases of 400 V and 230 V between any phase and the neutral star point (Figure 5.13).

The relationship between the phase voltage and the line voltage for a star connected winding is:

$$\text{Phase voltage} = \frac{\text{Line voltage}}{\sqrt{3}}$$

$$\text{Line voltage} = \text{Phase voltage} \times \sqrt{3}$$

Example:
The phase voltage is 230V for most domestic premises. The line voltage for this supply is

$$230 \times \sqrt{3} \quad = 230 \times 1.73$$
$$= 400 \text{ V}$$

Similarly a 400 V line voltage gives a phase voltage of:

$$\frac{400}{\sqrt{3}} \quad = \frac{400}{1.73} = 230 \text{ V}$$

Try this
For star connected windings
(a) calculate the phase voltages if the line voltages are:

240   _____

425   _____

300   _____

(b) calculate the line voltages if the phase voltages are:

415   _____

230   _____

110   _____

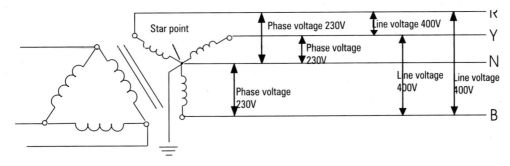

Figure 5.13   Voltages available from a star connected transformer winding

Load currents in three-phase circuits

It is important to recognise the relationships of the currents in star and delta connected windings. In star connected the current through the line conductors is equal to that flowing through the phase windings, as shown in Figure 5.14.

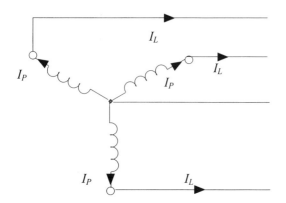

Figure 5.14

However in the delta connected winding this appears to be more complex. The line current, when reaching the transformer winding, is split into two directions so that two phase windings are each taking some current. As each of the phases are 120° out of phase with each other and the current is alternating each line conductor acts as a flow and return.

The arrows shown on Figure 5.15 only give an indication as to the current distribution from each line and all of these currents would not be flowing in the directions shown at the same time.

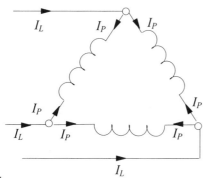

Figure 5.15

The current through the phase windings are:

$$I_P = \frac{I_L}{\sqrt{3}}$$

or $\quad I_L = I_P \times \sqrt{3}$

Example:

If the line current is 100 A the phase current is

$$I_P = \frac{100}{\sqrt{3}} = \frac{100}{1.73}$$
$$= 57.8\text{A}$$

Remember

For delta: $\quad U_L = U_P$

$$I_L = \sqrt{3} \times I_P \qquad \text{or } I_P = \frac{I_L}{\sqrt{3}}$$

For star: $\quad I_L = I_P$

$$U_L = \sqrt{3} \times U_P \qquad or\, U_P = \frac{U_L}{\sqrt{3}}$$

Try this

For delta connected windings
(a) calculate the phase currents if the line currents are:

100 A   _____
240 A   _____
2000 A   _____

(b) calculate the line currents if the phase currents are:

50 A   _____
240 A   _____
1 kA   _____

Points to remember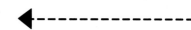

Distribution/star and delta circuits

The standard distribution voltages for the following are:

Heavy industry _____ or _____

Light industry _____ or _____

Commercial/domestic _____

The voltage between any two phases is _____, and the voltage between any phase and the neutral is _____.

On star connected windings the phase voltage = _____, and the phase current = _____.

On delta connected windings the phase voltage = _____, and the phase current = _____

The types of supply (4) which can be obtained from the secondary side of a delta/star transformer are:

Part 3

Three-phase power

In a three-phase system the power in each phase is the product of the phase voltage, phase current and power factor.

i.e. $U_P \times I_P \times \cos \theta$

therefore in a balanced system:

Total power $= 3 \times U_P \times I_P \times \cos \theta$

Note: Three-phase motors normally provide a balanced load.

Star connection

Power $= 3 \times U_P \times I_P \times \cos \theta$

But $U_P = \dfrac{U_L}{\sqrt{3}}$ and $I_P = I_L$

∴ Three-phase power $= \dfrac{3 U_L I_L \cos \theta}{\sqrt{3}}$

$= \sqrt{3} \, U_L I_L \cos \theta$

Delta connection

Power $= 3 U_P I_P \cos \theta$

But $U_P = U_L$ and $I_P = \dfrac{I_L}{\sqrt{3}}$

∴ Three-phase power $= \dfrac{3 U_L I_L \cos \theta}{\sqrt{3}}$

$= \sqrt{3} \, U_L I_L \cos \theta$

Remember

Three-phase power of a "balanced system" (star or delta connected) is calculated by the formula:

3-phase power $= \sqrt{3} \, U_L I_L \cos \theta$

Example

Three identical 20 Ω impedances, each with a power factor of 0.85 lagging, are connected to a 400 V 3-phase supply. Calculate the power dissipated when the impedances are connected in:

(a) star
(b) delta

(a) $U_P = \dfrac{U_L}{\sqrt{3}}$

$= \dfrac{400}{1.732}$

$= 231 \text{V}$

$I_L = I_P = \dfrac{U_P}{Z_P} = \dfrac{231}{20} = 11.55 \text{ A}$

$P = \sqrt{3} \, U_L I_L \cos \theta$

$= 1.732 \times 400 \times 11.55 \times 0.85$

$= 6801.564 \text{ W } (6.8 \text{ kW})$

OR total power

$= 3 \times U_P \times I_P \times \cos \theta$

$= 3 \times 231 \times 11.55 \times 0.85$

$= 6803.5275 \text{ W } (6.8 \text{ kW})$

(b) $I_P = \dfrac{U_P}{Z_P} = \dfrac{400}{20}$ (since $U_P = U_L$)

$= 20 \text{ A}$

$I_L = \sqrt{3} \, I_P$

$= 1.732 \times 20 = 34.64 \text{A}$

$P = \sqrt{3} \, U_L I_L \cos \theta$

$= 1.732 \times 400 \times 34.64 \times 0.85$

$= 20398.8 \text{ W } (20.4 \text{ kW})$

OR total power

$= 3 \times U_P \times I_P \times \cos \theta$

$= 3 \times 400 \times 20 \times 0.85$

$= 20400 \text{ W } (20.4 \text{ kW})$

Note: the power dissipated in delta is three times the power dissipated in star. (i.e. $3 \times 6.8 = 20.4$ kW)

Try this
(a) If the impedances shown each have a power factor of 0.9, calculate the total power dissipated.
(b) If the same impedances are now connected in delta, determine the total power dissipation.

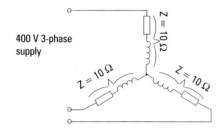

400 V 3-phase supply

$Z = 10\,\Omega$

$Z = 10\,\Omega$

$Z = 10\,\Omega$

Figure 5.16

Try this
For the circuit shown determine:
(a) the line current
(b) the phase current
(c) the total power dissipated

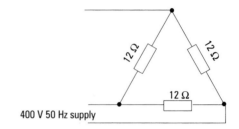

$12\,\Omega$ $12\,\Omega$

$12\,\Omega$

400 V 50 Hz supply

Figure 5.17

Example
Three 30 Ω resistors are connected in star to a 400 V, 3-phase, 4-wire a.c. supply.

Calculate
(a) the phase and line currents
(b) the total power dissipated

(a) $U_p = \dfrac{U_L}{\sqrt{3}} = \dfrac{400}{1.732} = 231\,\text{V}$

$I_P = \dfrac{U_P}{R_P} = \dfrac{231}{30} = 7.7\,\text{A}$

In star $I_L = I_P = 7.7\,\text{A}$

(b) $P = \sqrt{3}\,U_L\,I_L\,\cos\theta$

$= 1.732 \times 400 \times 7.7 \times 1$

$= 5334.56\,\text{W}\ (5.3\text{kW})$

OR total power $= 3 \times U_P I_P\,\cos\theta$

$= 3 \times 231 \times 7.7 \times 1$

$= 5336.1\,\text{W}\ (5.3\ \text{kW})$

Remember for purely resistive a.c. loads the power factor is unity, therefore $\cos\theta = 1$.

Example
A 30 kW, 400 V balanced 3-phase delta connected load has a power factor of 0.86 lagging. Calculate
(a) the line current
(b) the phase current

(a) $P = \sqrt{3}\,U_L\,I_L\,\cos\theta$

$\therefore I_L = \dfrac{P}{\sqrt{3}\,U_L\,\cos\theta} = \dfrac{30 \times 10^3}{1.732 \times 400 \times 0.86}$

$= 50.35\,\text{A}$

(b) $I_P = \dfrac{I_L}{\sqrt{3}} = \dfrac{50.35}{1.732}$

$= 29\,\text{A}$

Try this

A 24 kW electric furnace has three resistive elements connected in delta to a 400 V 3-phase supply. Calculate the resistance of EACH element.

Three-phase balanced loads

All transmission and distribution is carried out using a three-phase system. It is important that each of the phases carries about the same amount of current.

Three-phase motors have windings where each phase is the same and therefore the conductors carry the same current. These automatically create a balanced situation.

For domestic areas the output of the star connected transformer is 400/230V. All premises are supplied with a phase and neutral of 230V. To try and balance the loads on each of the phases the houses may be connected as shown in Figure 5.18.

Remember

It is always important to balance the loads across the phases of a three-phase supply.

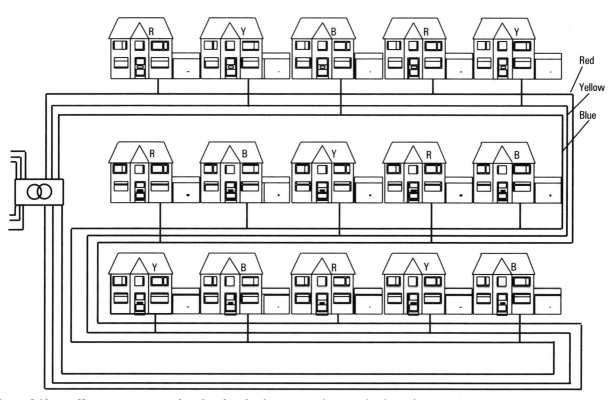

Figure 5.18   Houses are connected so that their loads are spread across the three phases

If it was possible to load all of the phases exactly the same the current in the neutral would be zero.

This can be proved by drawing the loads to scale on a phasor diagram, as shown in Figure 5.19.

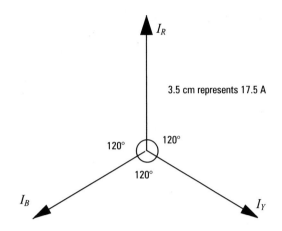

3.5 cm represents 17.5 A

Figure 5.19   Scale: 1 cm represents 5 A

Each line is drawn the same length to a scale equalling the current and each line is at equal angles (120°) to the one before, going in a clockwise direction.

By now "adding" the load of each phase to the other in the direction of the arrow and at the correct angle the resultant current I_N can be found.

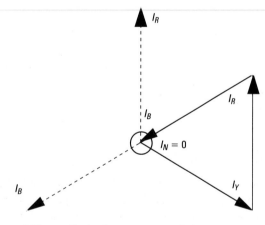

Figure 5.20   Scale: 1 cm represents 5 A

If the current in the phases are not equal then the resultant I_N would not be zero.

I_R = 150 A

I_Y = 200 A

I_B = 100A

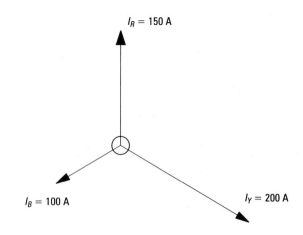

Figure 5.21   Scale: 1 cm represents 50 A

The current in the neutral I_N can be found by measuring the distance from the star point to.I_Y

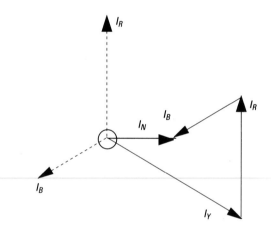

Figure 5.22   Scale: 1 cm represents 50 A

In this case it is

I_N = 1.7 cm
I_N = 85 A

This means that due to the phases not being balanced 85 A is flowing in the neutral conductor.

Reasons for balancing single-phase loads on a three-phase system.

- to limit the current in the neutral conductor (I_N)

- to prevent excessive loading on any one phase conductor (R, Y or B), thus keeping the size of the cable used to a minimum

- to reduce the voltage drop in the system
  (If $I_N = 0$ then $I_N \times R_N$ will be zero)

- to reduce the power loss in the system
  (If $I_N = 0$ then $I_N^2 \times R_N$ will be zero)

Try this

The currents on a 3 phase star connected load are

I_R = 25 A
I_Y = 15 A
I_B = 10 A

Draw a scaled phasor diagram and determine the current in the neutral.

Points to remember ◀ - - - - - - - - - -

3-phase power/3-phase balanced loads

Three-phase power of a "balanced system" (_____ or _____ connected) can be calculated by the formula: _____,

or the formula: total power = _____

Single-phase loads should be balanced on a three-phase system mainly:
1. to _____ the current in the _____ conductor
2. to prevent any one phase taking _____
   _____ current

On a balanced three-phase system the _____ in the neutral would be _____.

The current in the neutral conductor can be found by constructing a _____ _____ to scale.

Part 4

Interacting fields

The turning effect, or torque, produced by a.c. induction motors, either three-phase or single-phase, depends upon the interaction of two magnetic fields. One of these fields is produced by the "stator" windings, which carry current from the supply, and the other field is produced by the "rotor" windings or "rotor" conductors, when a current is induced into them.

When the a.c. supply is connected to the stator windings of a three-phase induction motor a rotating-magnetic-field is set up by these windings.

Production of a rotating magnetic field from a three-phase supply

The three-phase stator windings may be simplified by regarding them as three coils set at 120 degrees (Figure 5.23).

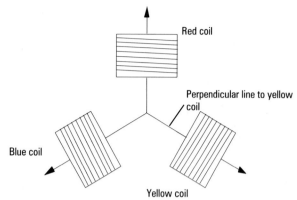

Figure 5.23   *Simplified end view of stator coils*

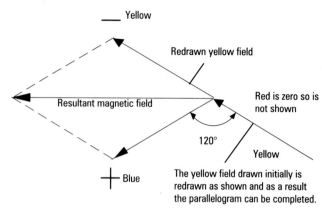

Figure 5.25

The magnetic field produced by each coil will change direction, but its magnetic force will act along the line drawn perpendicular to each coil, irrespective of the direction in which the lines of force travel.

Now assume the three coils are connected to a three-phase supply, the waveforms of which are shown in Figure 5.24.

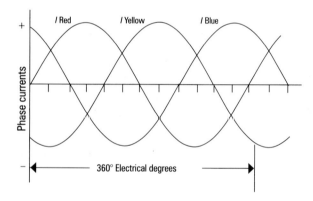

Figure 5.24   *Phase current waveforms*

Let's consider the arrangements of coils in Figure 5.23 at the point when the red phase is at 0°.

I_{Red}	= 0 amps
I_{Yellow}	= some value in a negative direction
I_{Blue}	= the same value in a positive direction

A diagram representing the magnetic forces which exist in the area enclosed by the coils can be drawn; it is a parallelogram of forces acting at a specific time.

Note: A current in a positive direction will produce a field indicated by an arrow pointing away from the centre of the coil arrangement in Figure 5.23, and a current in a negative direction produces a field represented by an arrow pointing towards the centre of the coil arrangement.

Constructing a parallelogram

Example as before with red phase at 0°.

Draw two lines, equal in length, displaced 60°.
(Refer to Figures 5.23 and 5.24 to determine field directions.)

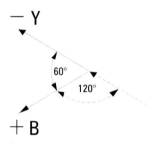

Figure 5.26a

Next draw two dotted lines, parallel to the first two lines and equal in length, to complete the parallelogram.

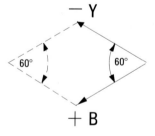

Figure 5.26b

Finally draw in the intersecting line to show the resultant field magnitude and direction.

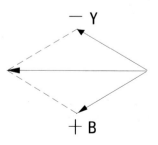

Figure 5.26c

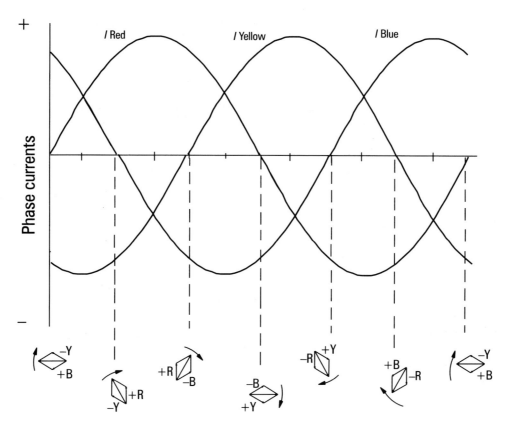

Figure 5.27   Parallelograms of forces

At 0°	At 60°	At 120°	At 180°	At 240°	At 300°	At 360°
$I_R = 0$	$I_R = +$	$I_R = +$	$I_R = 0$	$I_R = -$	$I_R = -$	$I_R = 0$
$I_Y = -$	$I_Y = -$	$I_Y = 0$	$I_Y = +$	$I_Y = +$	$I_Y = 0$	$I_Y = -$
$I_B = +$	$I_B = 0$	$I_B = -$	$I_B = -$	$I_B = 0$	$I_B = +$	$I_B = +$

The parallelogram shown is redrawn at 60° intervals to illustrate how the magnetic forces and fields change directions as the phase currents change.

The magnetic field has now rotated one complete revolution and its speed will depend upon the frequency of the supply and the number of pairs of poles on the motor.

The speed at which the rotating magnetic field travels is termed the "synchronous speed".

Percentage voltage drops throughout an installation

The maximum permissible voltage drop (in compliance with I.E.E. Wiring Regulation 525-01-02) is 4% of the nominal voltage of the supply between the origin of the installation (usually the supply terminals) and a socket outlet or the terminals of fixed current-using equipment (for example motor, cooker or fluorescent luminaire).

Example

A small industrial premises is supplied with a 400 V 3-phase 50 Hz supply (Figure 5.28).

(a) Calculate the percentage voltage drop for EACH of the following parts of this installation.
  (i) the ground floor distribution board where the line voltage measured is 396 V.
  (ii) the first floor distribution board where the line voltage measured is 390 V.
  (iii) the motor final circuit no. 2 where the line voltage measured is 385 V.
  (iv) the furthest lighting point where the phase voltage measured is 222 V.
(b) Do the percentage voltage drops (in (a) above) comply with BS 7671?

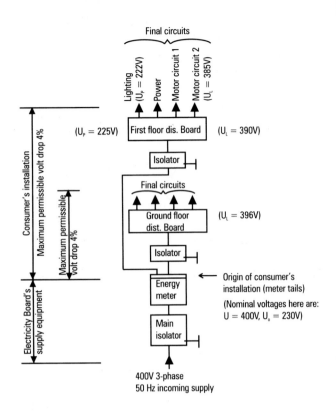

Figure 5.28   *Small industrial premises*

Answer

(a)

(i) Percentage voltage $= \dfrac{396}{400} \times 100 = 99\,\%$

∴ Percentage volt drop $= 100 - 99 = 1\%$

(ii) Percentage voltage $= \dfrac{390}{400} \times 100 = 97.5\%$

∴ Percentage volt drop $= 100 - 97.5 = 2.5\%$

(iii) Percentage voltage $= \dfrac{385}{400} \times 100 = 96.25\%$

∴ Percentage volt drop $= 100 - 96.25 = 3.75\%$

(iv) Percentage voltage $= \dfrac{222}{230} \times 100 = 96.52\%$

∴ Percentage volt drop $= 100 - 96.52 = 3.48\%$

(b) Yes, since they are less than 4%.

Note: The maximum permissible voltage drop values are:

(i) for 230 V single-phase circuits $= 230 \times \dfrac{4}{100} = 9.2\,\text{V}$

(ii) for 400 V three-phase circuits $= 400 \times \dfrac{4}{100} = 16\,\text{V}$

D.C. supplies from an a.c. source

A.C. to d.c. converters

There are several different circuits that can be used to produce a direct current from an a.c. source. Usually the circuit depends on the degree of smoothness of the d.c. that is required. If the d.c. is not very critical then a simple half wave circuit can be used as shown in Figure 5.29. It has been assumed in all of the diagrams that the d.c. voltage is less than the a.c. and has to be transformed down.

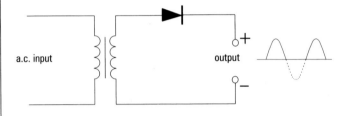

Figure 5.29   *Half-wave rectifier circuit*

This can be made a smoother output by placing an electrolytic capacitor across the load terminals.

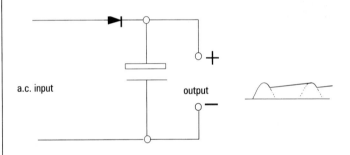

Figure 5.30

Where a better output is required a full wave circuit is used. This may use a full wave bridge circuit or a transformer with a centre tap on the output and a bi-phase circuit.

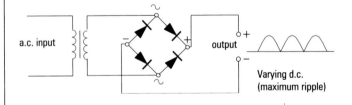

Figure 5.31   *Full-wave bridge rectifier circuit*

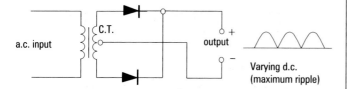

Figure 5.32   *Bi-phase (full-wave) rectifier circuit*

Sometimes it is necessary to be able to select different voltages of d.c. In this case the secondary tappings and a selector switch will need to be incorporated in the low voltage a.c. part of the circuit, as shown in the example in Figure 5.33.

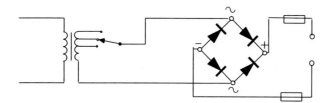

Figure 5.33

These circuits can be smoothed out if required by using some form of filter circuit. Examples are shown in Figures 5.34 and 5.35.

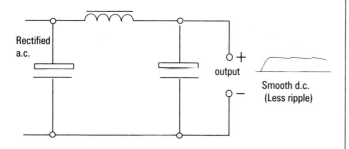

Figure 5.34   *L.C. smoothing filter*
             Commonly known as π *(pi) type*

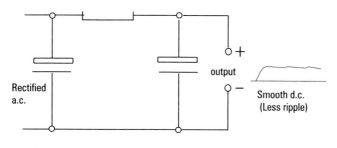

Figure 5.35   *R.C. smoothing filter*

The above circuits are ideal if the output is not critical for in most cases when the circuit is loaded the voltage will drop. To overcome this a voltage regulator can be included in the circuit as the shown in the example in Figure 5.36.

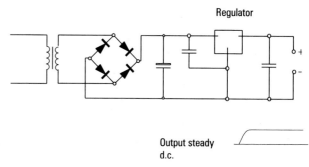

Figure 5.36   *Regulated a.c./d.c. power supply*

Voltage stabilisation

The d.c. output voltage can be stabilized by employing a zener diode. (Figure 5.37)

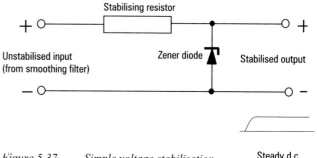

Figure 5.37   *Simple voltage stabilisation*
             circuit

> ### Remember
> A rectifier converts a.c. to d.c.

Rotating magnetic fields/percentage voltage drops/a.c. to d.c. supplies

The torque produced by an a.c. induction motor depends upon the _____ of two _____ fields. One of these fields is produced by the _____ windings, and the other field is produced by the _____ windings.

When the a.c. supply is connected to the _____ windings of a 3-phase induction motor a _____ _____ _____ is set up by these windings.

The speed of the rotating-magnetic-field depends upon the _____ of the supply and the number of _____ ___ _____ on the motor.

The maximum permissible voltage drop is _____ of the_____ voltage of the supply between the _____ of the installation and the_____ of an a.c. induction motor.

A rectifier converts _____ to _____.

Half-wave rectifiers have _____ diode.

Full-wave rectifiers have _____ diodes when supplied by a transformer with a _____ _____ secondary winding.

A full-wave bridge rectifier has _____ diodes.

L.C. and R.C. filters are used to _____ the output waveform of a rectifier.

The d.c. output voltage can be stabilized by using a _____ _____ .

Try this

The diagram shown is of a typical a.c./d.c. power supply incorporating smoothing and stabilisation.
(a) State the function of each stage 1, 2, 3 & 4.
(b) Sketch typical waveforms for each part of the circuit A, B, C, D and E.

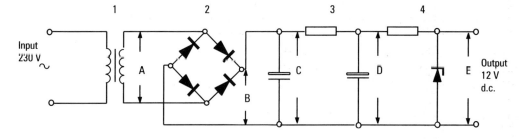

Figure 5.38

Self assessment short answer questions

1. Fill in the voltages that are most likely to be encountered in the following situations:

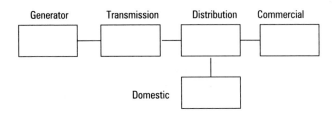

Figure 5.39

2. Complete the missing values in the table below for both star and delta arrangements.
The first one is completed as an example.

Star

U_P	U_L	I_P	I_L
240 V	415 V	5 A	5 A
110 V		25 A	
	380 V		15 A
	240 V		100 A
680 V		125 A	

Delta

U_P	U_L	I_P	I_L
415 V	415 V	5 A	8.65 A
110 V		25 A	
	380 V		15 A
	240 V		100 A
680 V		125 A	

3. Draw a scaled phasor diagram and calculate the current in the neutral conductor if the resistive phase loads are as follows:

        Red phase   50 A

        Yellow phase   70 A

        Blue phase   35 A

State the scale you are working to.

4. A small factory has a three-phase load of 56 kW at 60 kVA$_r$. Determine:
(a) the kVA
(b) the power factor

5. Draw a circuit diagram to show how a 24 V d.c. supply may be obtained from a 230 V a.c. supply using a centre tapped transformer.

7. (a) At what speed does a rotating magnetic field travel?
   (b) State two factors that the speed of the rotating magnetic field depends upon.
   (c) Complete the diagram shown below to indicate the resultant magnetic fields magnitude and direction.

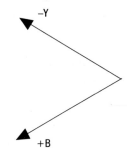

Figure 5.40

6. A small industrial unit is supplied at 400 V, 50 Hz. Calculate the percentage voltage drop for EACH of the following parts of the installation.
   (a) A ground floor distribution board where the voltage is measured as 395 V.
   (b) A second floor distribution board where the voltage is measured as 388 V.
   (c) A motor final circuit having a voltage of 382 V.

8. Draw a circuit diagram and on it show how EACH of the following supplies can be obtained from the secondary side of a delta/star transformer:
   (a) 400 V 3-phase 4 wire
   (b) 400 V 3-phase 3 wire
   (c) 400 V single-phase
   (d) 230 V single-phase

Progress Check

1. (a) State the unit for
   (i) magnetic flux

   (ii) magnetic flux density

   (b) State the symbol used for EACH quantity in (a) above.

2. A copper conductor 0.2 m long is situated in and at right angles to a magnetic field and experiences a force of 4 N. Calculate the magnetic flux density if the current in the conductor is 20 A.

3. Three capacitors of 8, 12 and 24 µF respectively are connected in series. If the supply voltage is 200 V, calculate:
   (a) the total capacitance
   (b) the charge on each capacitor

4. State:
   (a) three factors that affect the capacitance of a capacitor
   (b) three dielectric materials used in capacitors

5. Figure 5.41 is the circuit diagram for a particular type of d.c. motor. Identify the component parts labelled A, B and C.

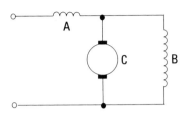

Figure 5.41

6. Explain briefly why starting resistances are connected in series with the armature in a d.c. shunt motor circuit.

7. Sketch an appropriate waveform diagram for EACH of the a.c. circuits shown in Figures 5.42a, 5.42b and 5.42c.

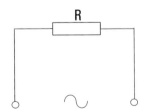

Figure 5.42a

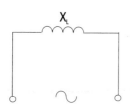

Figure 5.42b

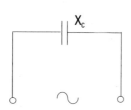

Figure 5.42c

8. (a) A 230 V fluorescent luminaire dissipates 125 W. Calculate the current taken from the supply if the power factor is:
    (i) 0.8 lagging
    (ii) unity
    (b) State how the power factor of the luminaire may be improved.

9. For the circuit shown in Figure 5.43 below calculate the
    (a) line current
    (b) phase current
    (c) power dissipated

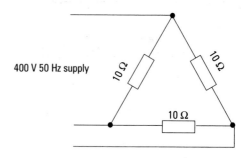

Figure 5.43

10. Draw a circuit diagram showing how a 24 V d.c. supply can be obtained from a 230 V, 50 Hz a.c. supply using a rectifier consisting of two diodes.

6

A.C. Machines

Answer the following questions to remind yourself of what was covered in Chapter 5.

1. For the circuit shown below calculate:
   (a) the line and phase currents
   (b) the power dissipated

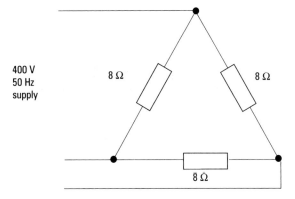

400 V
50 Hz
supply

8 Ω   8 Ω

8 Ω

Figure 6.1

2. Draw a phasor diagram, to a scale of 1 cm = 10 A, for a star connected three-phase balanced circuit with 45 A per phase.

3. (a) Sketch circuit diagrams of:
   (i) an RC smoothing filter
   (ii) an LC smoothing filter

   (b) Draw a typical waveform representing the output of these types of filter.

4. Draw the output waveform of a three-phase generator, indicating the phase displacement.

On completion of this chapter you should be able to:

◆ explain with the aid of diagrams the production of torque from a three-phase supply
◆ state the relationship between motor speed, frequency and the number of poles
◆ determine slip, synchronous speed and rotor speed for a.c. motors
◆ describe with the aid of diagrams the construction of three-phase and single-phase a.c. motors
◆ describe the basic operation of three-phase and single-phase a.c. motors
◆ describe with the aid of diagrams starting methods for three-phase and single-phase a.c. motors
◆ state the purpose of the following devices in motor circuits:
   (a) overload protection
   (b) no volt/undervoltage protection
   (c) thermistor protection
   (d) emergency stop buttons
   (e) remote start/stop control
◆ identify the different types of motor enclosure

Part 1

Production of torque from a three-phase supply

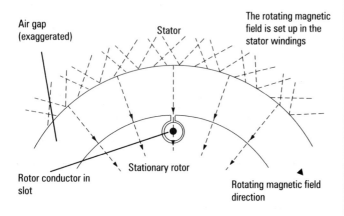

Figure 6.2

Consider one conductor on the stationary rotor (Figure 6.2) and the three-phase supply to the stator producing a clockwise rotating magnetic field.

This rotating magnetic field sweeps across the air gap to the rotor and also across the stationary conductor. As it does so a current is induced into the conductor which flows towards you, and sets up a concentric anticlockwise field round the conductor. (Figure 6.3)

The direction in which the current is induced can be determined by applying Fleming's Right Hand Rule.

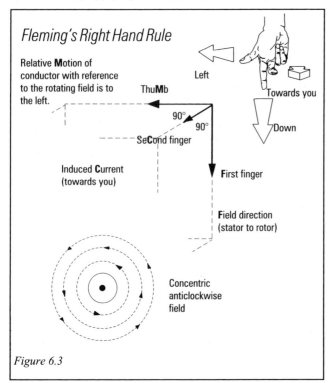

Figure 6.3

These two fields interact with each other causing a strengthening of the field to the left of the conductor and a weakening of the field to the right. (Figure 6.4) Thus a force will be applied to the conductor in the same direction as the rotating magnetic field. (Apply Fleming's Left Hand Rule.)

Similar forces are applied to all the rotor conductors, so that a torque is produced causing the rotor to rotate in the same direction as the rotating field.

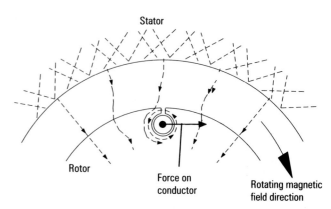

Figure 6.4

Reversal of rotation

The standard sequence of a three-phase supply is red, yellow and blue or L_1, L_2 and L_3.

The direction in which the magnetic field rotates is directly related to the sequence that the phases are connected to the stator windings. Rotation of the field can be reversed by reversing the connection of any two incoming phases. Consequently the rotation of the rotor will also be reversed.

Remember
The rotor rotates in the same direction as the rotating magnetic field.

The direction of the rotating field depends upon the sequence of the three phases.

Reverse any two phases to reverse the direction of the rotating field.

Slip

We have seen that the relative motion between the conductors and the rotating magnetic field creates a torque. This produces rotation and the rotor increases speed in the same direction as the rotating field.

If the rotor reached the synchronous speed of the rotating field then the conductors would be moving with the field, there would be no relative movement between conductors and field, no current would be induced , so there would be no torque produced. Thus it is not possible for a normal induction motor to run at synchronous speed.

If torque is to be produced the induced current is necessary therefore the rotor must run at something less than the synchronous speed of the rotating field.

The difference between the actual rotor speed (N_r) and synchronous speed (N_s) is called the "slip speed".

It is more usual to speak of "slip" of an induction motor as a fractional (per unit) value which is given the symbol s.

$$s = \frac{N_s - N_r}{N_s}$$

or as a percentage value:

$$s(\%) = \frac{N_s - N_r}{N_s} \times 100$$

Remember
There must be relative movement between rotor conductors and the rotating magnetic field for torque to be produced.

Speed, poles and frequency (three-phase winding)

A three-phase stator winding consists of three sets of coils evenly distributed around the core of the stator and connected in either "Star" (Y) or "Delta" (Δ). Each of these windings could have two or more poles per phase depending upon the speed required.

The rotating magnetic field set up by a two pole winding completes one revolution (360°) in one complete cycle of the mains supply, whilst with a four pole winding (2 pairs of poles) it completes one revolution in two cycles.

It should be noted then:

as the number of poles per winding is increased so the speed of the rotating magnetic field within the machine decreases (see Table 6.1).

Table 6.1

Synchronous speeds and standard rotor speeds at full load (50 Hz supply)

Poles	Synchronous speeds revs/min	Rotor speeds revs/min
2	3000	2900
4	1500	1440
6	1000	960
8	750	720
10	600	580
12	500	480
16	375	360

Synchronous speed
As previously mentioned the "synchronous speed" of a.c. motors depends on
- the frequency of the supply
- the number of pairs of poles on the stator

Synchronous speed can be calculated by using

$$N_s = \frac{f}{p} \times 60$$

Where:

N_s is the synchronous speed in rev/min

f is the frequency in Hertz (cycles per second)

p is the number of pairs of poles

(60 converts frequency to cycles per minute so that the speed can be calculated in rev/min)

Or use

$$n_s = \frac{f}{p}$$

where n_s is the synchronous speed in revolutions per second (rev/s)

Speed and slip calculations

Example
We need to calculate the synchronous speed in rev/min of a 16 pole motor if the supply frequency is 50 Hz.

$$N_s = \frac{f}{p} \times 60 = \frac{50}{8} \times 60 = 375 \text{ rev/min}$$

Remember p = number of pairs of poles in these calculations.

Try this

Determine the synchronous speed in rev/s of a 4 pole motor connected to a 50 Hz supply.

Example

If a twelve pole, 50 Hz induction motor runs at 475 rev/min we can calculate
  (a) the synchronous speed
  (b) the percentage slip

(a) $N_s = \dfrac{f}{p} \times 60 = \dfrac{50}{6} \times 60$

   $= 500$ rev/min

(b) $s(\%)\dfrac{N_s - N_r}{N_s} \times 100 = \dfrac{500 - 475}{500} \times 100$

   $= \dfrac{25}{500} \times 100$
   $= 0.05 \times 100$
   $= 5\%$

Try this

If an eight pole induction motor runs at 12 rev/s and is supplied from a 50 Hz supply calculate the percentage slip.

Example

It has been established that an induction motor has 6 poles and a per unit slip of 0.05 at full load when the supply frequency is 50Hz. Before we can determine the rotor speed of the motor we must first calculate the synchronous speed.

$$n_s = \frac{f}{p} = \frac{50}{3} = 16.67 \text{ rev/s}$$

Now using the transposition of $s = \dfrac{n_s - n_r}{n_s}$ so that

$n_r = n_s(1 - s)$
  $= 16.67 (1 - 0.05)$
  $= 16.67 \times 0.95$
  $= 15.84$ rev/s

Try this

Calculate:
(a) the synchronous speed and
(b) the rotor speed of a 2 pole induction motor which has a slip of 0.03 per unit when operating at full load on a 50 Hz supply.

Points to remember ◀ ─ ─ ─ ─ ─ ─ ─ ─ ─ ─

Torque/slip/speed/poles and frequency

For a torque to be produced, there must be a relative movement between the _____conductors and the _____
_____.

Slip speed is the difference between the _____speed and the _____speed.

The synchronous speed of a.c. motors depends on:
1.

2.

What do the following represent in the formula
$n_r = n_s(1 - s)$?

   n_r is
   n_s is
   s is

Part 2

The construction and operation of three-phase a.c. motors

Three-phase motors are widely used for industrial drives. When they are compared with single-phase motors for similar loads they are

- less expensive
- of smaller dimension
- more efficient and
- generally self starting.

Figure 6.5   Three-phase induction motor

The current taken by a three-phase motor is between one quarter and one third of that taken by a single-phase motor with a similar power rating. This can be seen in Table 6.2.

Table 6.2 Current for a.c. motors

This table shows a (rule of thumb) current per phase taken by modern induction motors (speed 1440 rev/min) allowing reasonable efficiencies and power factor. When making calculations for real refer to manufacturers' details for precise data.

h.p.	¼	½	¾	1	2	3	5½	7½	10	15	20	25
kW rating	0.18	0.37	0.55	0.75	1.5	2.2	4.0	5.5	7.5	11.0	15.0	18.5
230V single phase	2.7	4.2	6.3	7.7	12.0	14.4	24.0	32.0	40.0	56.0	76.0	94.0
400V three phase	0.8	1.2	1.5	2.0	3.6	4.9	8.3	11.0	15.0	21.0	28.0	35.0

The power rating of a motor is measured in watts, or more usually, kilowatts. Motor power was formerly measured in horsepower.

1 h.p. = 746 watts or 0.75 kW

Remember

3-phase motors are:

- installed where a larger amount of power is required
- cheaper and more efficient than single-phase
- generally self starting

1 h.p. = ¾ kW

Try this

Determine the per unit slip for a motor as shown in Table 6.2 with a speed of 1440 rev/min if the supply frequency is 50 Hz.

Construction of three-phase cage rotor induction motors (formerly called Squirrel Cage Motors)

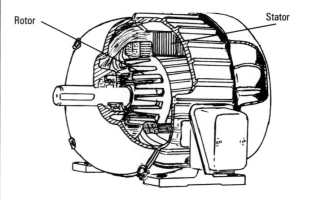

Rotor   Stator

Figure 6.6   Three-phase induction motor

This motor has a fixed part, called a "stator", which houses the three-phase windings that produce the rotating magnetic field, and a moving part called a "cage-rotor" which revolves within the stator.

The stator core is built up from steel laminations with slots to receive the windings The laminations are punched from sheet steel about 0.5mm thick, and are lightly insulated on one or both sides. Large stators are built up with segmental stampings (Figure 6.7).

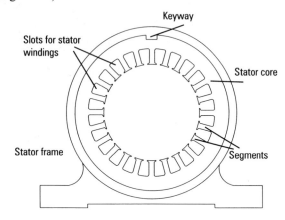

Figure 6.7   *Segmental stator core*

The rotor core is also built up from steel laminations, having longitudinal slots into which lightly insulated copper or aluminium conductors (called "rotor bars") are fitted. The rotor bars are short circuited at each end by a heavy copper or aluminium ring so forming a closed circuit (Figure 6.8).

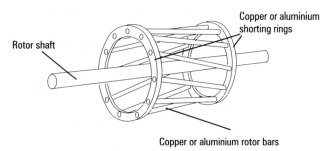

Figure 6.8   *Cage-rotor*

The rotor bars are "skewed" (at an angle) to the end rings. This type of design reduces magnetic noise (hum) and crawling tendencies during starting and run up.

Remember
- The fixed part is the stator and the moving part is the rotor.
- The stator and rotor cores are made from steel laminations.
- The rotating magnetic field is produced by the stator windings.
- The rotor bars are fitted into slots in the rotor core and are short-circuited by end rings.

Laminated cores

Stator and rotor cores, like transformer cores, are laminated to reduce "eddy currents". These are circulating currents which are induced into the core when a magnetic flux cuts through the core. These currents cause unnecessary heating and power losses in the core, i.e. the larger the current the more the heat dissipation (Figure 6.9).

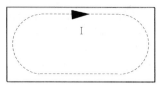

Figure 6.9   *Large solid core*

- Larger cross-sectional area – lower resistance path
- Large circulating current – larger amount of heat dissipated

Figure 6.10   *Core made up of 4 laminations insulated from each other.*

- Each lamination has a smaller cross-sectional area – resistance path is higher.
- Circulating current in each lamination is smaller – smaller amount of heat dissipated.

The power loss is best explained with the following example.

Example

If the two cores in Figures 6.9 and 6.10 had the same resistance (say 4 Ω) and the circulating current was say 4 A then the power loss would be

a) solid core

$$P = I^2 \times R = 16 \times 4 = 64 \text{ W}$$

b) laminated core

The loss in each lamination is
$$P = I^2 \times R = 1^2 \times 1 = 1 \text{ W}$$
for the whole core it is 4 W

Remember
Stator and rotor cores are laminated to reduce power losses due to "eddy currents".

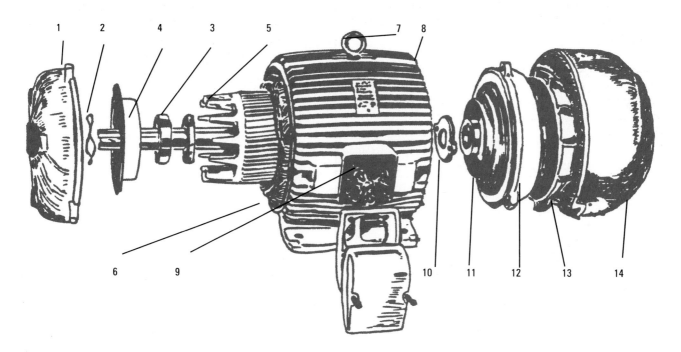

Figure 6.11   Component parts of a typical cage-rotor induction motor

Component parts

1. End shield, drive end
   (welded or rolled steel)
2. Thrust washer
3. Ball bearing, drive end

4. Flume

5. Cage rotor (cast aluminium cage)

6. Stator windings

7. Eyebolt

8. Stator frame
   (welded or rolled steel)

9. Terminal box

10. Inside bearing cap, non-drive end

11. Ball bearing, non-drive end
12. End shield, non-drive end (welded or rolled steel)
13. Fan
    (polypropylene or aluminium alloy)
14. Fan cover
    (aluminium or pressed steel)

Function of the component parts.

1. & 12. The endshields form part of the motor enclosure
   and support the rotor on its ball bearings.
2. The thrust washer takes up any float on the rotor.
3. & 11. The ball bearings support the rotor shaft and
   enable it to turn freely.
4. The flume protects the stator windings from
   mechanical damage.
5. The rotor supports the rotor bars and the end rings,
   and provides a magnetic path to help produce the
   driving torque.
6. The stator windings produce the rotating magnetic
   field.
7. The eyebolt is used as an anchorage point for a
   sling and shackle when lifting the motor off its
   bedplate.
8. The stator frame supports the stator core and
   windings and also provides a magnetic path for the
   rotating magnetic field.
9. The terminal box facilitates the connection of the
   supply to the stator windings.
10. The inside bearing cap prevents ingress of dust
    into the bearing and grease out into the motor.
11. As 3.
12. As 1.
13. The fan cools the motor by circulating air between
    the fins on the stator frame.
14. The fan cover guards the fan impeller.

Note: Motor enclosures (frame and endshields) are also made of cast iron or aluminium.

Connection of the three-phase cage-rotor induction motor

Let's take a look inside the terminal box of this three-phase cage-rotor motor.

There are six stud type terminals moulded into an insulated terminal board to which each end of the three phase windings are connected as shown below.

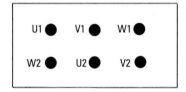

Figure 6.12   Terminal board

This enables the windings to be connected either in "star" or "delta" configuration.

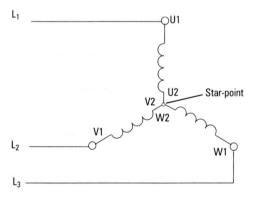

Figure 6.13   Star connection

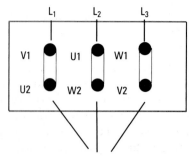

Three links required to form delta connection

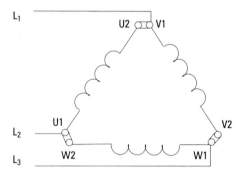

Figure 6.14   Delta connection

Many three-phase cage-rotor motors have just three terminations inside the terminal box. The windings in this case will be permanently connected in either "star" or "delta" configuration. It is not normally possible to change the connections as they are internally connected.

Note: Formerly motor terminal markings were A1, B1, C1 and A2, B2, C2 for a six terminal connection, and A, B, C for a three terminal connection, now replaced with U, V, W. Supply line connections L_1, L_2 and L_3 may have alternative markings.

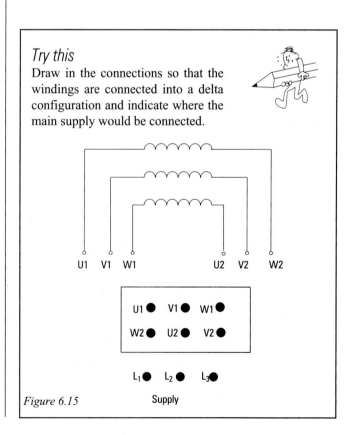

Operation of the three-phase cage-rotor induction motor

When the three-phase a.c. supply is connected to the stator windings a rotating magnetic field is set up by the windings. This rotating field cuts through the rotor bars and induces e.m.f.'s in them. Since the rotor bars are short circuited by end rings, considerable induced currents flow in them which in turn create their own magnetic fields. These fields interact with the stator's rotating magnetic field and exert magnetic forces upon the rotor bars. These forces cause the rotor to turn in the same direction as that of the rotating magnetic field.

The rotor never quite manages to catch up with the synchronous speed of the rotating magnetic field, if it did, then no e.m.f. and current would be induced in the rotor bars (remember in Part 1 we said that there must be a relative motion between the rotating field and the rotor bars to create a torque) and the rotor would tend to slow down and stop.

The rotor will settle down to a steady speed just below synchronous speed. This difference between synchronous speed and the actual rotor speed is called the "slip" speed, since the rotor tends to "slip" behind the rotating magnetic field of the stator.

Note: Induction motors are called "Asynchronous" motors since they do not run at synchronous speed.

Remember
To change direction of rotation change any two phase lines.

The rotating magnetic field is set up by the stator windings.

It is this rotating field which cuts through the bars on the rotor and induces e.m.f.'s into them.

The interacting fields (stator and rotor) cause the rotor to turn.

The rotor turns in the same direction as the rotating magnetic field.

The rotor speed is just below synchronous speed.

Advantages of a cage-rotor induction motor

- Due to the simple construction (only the stator winding is connected to the supply) the motor is comparatively cheap.
- Due to the cage construction of the rotor they are mechanically strong and robust, therefore they are particularly useful for industrial drives.
- They are normally self-starting (i.e. no special starting equipment is required).
- They require little maintenance as there are no rubbing contacts (brushes) on the rotor.
- Almost constant speed.

Disadvantages

- High starting current (up to 6 or 7 times full load current).
- Due to the inductance of the stator winding it has a low power factor (typical 0.7). This is even worse at the instant of starting.
- Poor starting torque and therefore they can only be started with light loading.

Typical applications:

- Machine tools
- Industrial drives
- Small pumps

Starting torque of cage-rotor motors

The starting torque depends on the design of the cage-rotor.

One main drawback with the cage-rotor motor is that it has a relatively high starting current and a low starting torque.

The "double-cage-rotor" motor has a lot better starting torque than the "single-cage-rotor" motor.

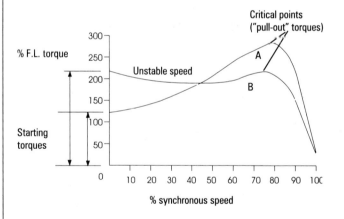

Figure 6.16   *Typical torque/speed characteristics curves for three-phase induction motors*

In Figure 6.16 above:

(A) with single-cage-rotor
(B) with double-cage-rotor

In each case the "knee" of the curve indicates the point where the motor "pulls-out" owing to the torque increasing above a certain value.

If the motor is loaded beyond this point it will no longer take the load and the speed will fall quickly to zero.

The double-cage-rotor motor

The "double-cage-rotor" motor is designed to provide a high
starting torque with a low starting current. The rotor is so
designed that the motor operates with the advantage of a high
resistance rotor circuit during starting (outer bars), and a low
resistance rotor circuit under running conditions (inner bars).
Figure 6.17.

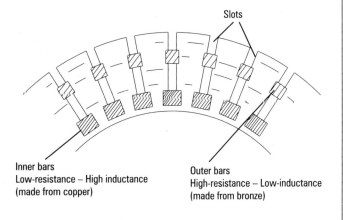

Slots

Inner bars
Low-resistance – High inductance
(made from copper)

Outer bars
High-resistance – Low-inductance
(made from bronze)

Figure 6.17   Section of rotor

Typical applications:

• Conveyors
• Industrial drives

Three-phase wound rotor induction motor

The stator construction is identical to that of the cage rotor
motor, as previously described.

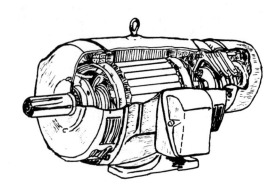

Figure 6.18   Three-phase wound rotor induction motor

The rotor consists of three windings having many turns
connected in "star" or "delta" configuration, the terminations
of which are brought out and connected to three slip rings fitted
on the rotor shaft. (Figure 6.19)

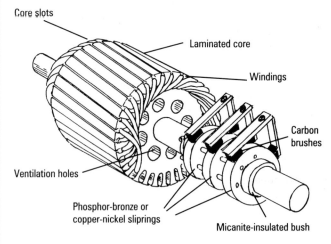

Core slots

Laminated core

Windings

Carbon
brushes

Ventilation holes

Phosphor-bronze or
copper-nickel sliprings

Micanite-insulated bush

Figure 6.19   Wound rotor with slip rings

The slip rings allow connection of the rotor windings to
external resistances as shown on Figure 6.20.

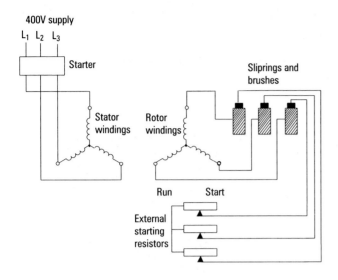

Figure 6.20

Operation

The motor is started with all the external resistance connected into the rotor circuit. Then the resistance is gradually reduced as the motor gains speed and at full speed is cut out altogether. The slip rings are then shorted out and the motor continues to run as a cage motor.

In large motors, which sometimes run almost continuously, a brush lifting device is incorporated with the shorting gear to eliminate needless brush and slip ring wear. The wound rotor, like the cage rotor has no external electrical connections to the supply.

For some applications the rotor slip rings are not shorted out and the rotor resistance is used as a form of speed control.

Typical applications:

* Large pumps
* Compressors

Figure 6.21   Drip proof wound-rotor induction motor
Sliprings are located beneath non-drive end cover

Changing the direction of rotation

The direction of rotation may be changed by changing any two phase lines.

Remember
The wound rotor windings are brought out to three slip-rings.

The slip-rings enable the rotor windings to be connected to external resistances.

The motor starts with all the resistance in the rotor circuit.

At full speed the resistance is cut out.

Points to remember ◀ – – – – – – – – – – – – – –

Construction and operation of 3-phase a.c. motors

A cage rotor has copper or _____ which are _____ by end rings. Stator and rotor cores are _____ to reduce _____.

When both ends of each winding of a 3-phase motor are brought out to the terminal box the motor can be connected in _____ or _____.

The 3-phase cage-rotor induction motor has a _____ starting current and a _____ starting torque.

The double-cage-rotor motor has a _____ starting current and a _____ starting torque than a single-cage-rotor motor.

The wound-rotor motor is started with all the _____ in the rotor circuit, and at full speed they are _____ _____.

Part 3

Operation of single-phase induction motors

As we have seen a three-phase induction motor is self starting due to the rotating magnetic field of the stator. This magnetic field cuts the rotor conductors and exerts a force on them which causes the rotor to turn.

However a simple single-phase induction motor with a single winding is not self starting, because if the single winding is fed with a.c., it simply produces a "pulsating field" which rises and falls with the alternating current, and this type of field produces no torque in the rotor.

Surprising as it may seem we can actually start this simple motor by spinning the rotor. The rotor conductors are still being cut by the pulsating field, but in addition, the conductors are moving through the field. The position now is that we have two fields (Rotor and Stator) interacting with each other, and the same conditions exist as in a rotating field. Once started the rotor will continue to run in that direction.

Obviously this arrangement is most unsatisfactory and a means of making single-phase motors "self starting" have to be considered. The basic principle being to get the stator field to move (rotate) so that the rotor may follow. Several methods are employed and all of them use an additional winding. This additional winding is called the "start" winding and it is fitted to the stator of the motor.

This start winding is usually wound with about the same number of turns as the main winding but of smaller diameter wire. The start winding is also not fitted so deep into the stator iron core as the main winding. This results in the start winding having a higher resistance and less inductive reactance than the main winding and is more in phase with the supply voltage. The thicker main winding has less resistance and a greater inductive reactance. The start winding is usually short-time rated and would overheat if left in the circuit for more than a few seconds.

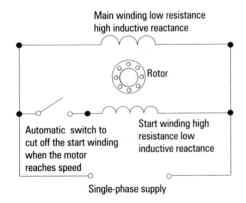

Figure 6.22   *Note: The Auto-switch is known as a centrifugal switch*

The magnetic fields produced by these out of phase currents create the necessary torque on the rotor to make the motor self starting when connected to the single-phase supply.

Two motors operating on this principle are called
* the split-phase motor and
* the capacitor split-phase motor
and the approximate phase displacement between the currents are 30° for the split-phase and 90° for the capacitor start split-phase.

Figure 6.23 shows how the rotating magnetic field of a capacitor-start, split-phase induction motor is produced.

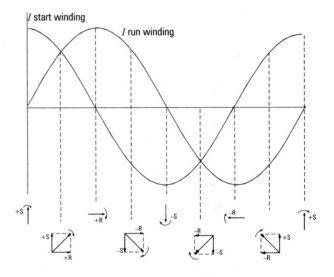

Figure 6.23   *Production of a rotating magnetic field from a single-phase capacitor-start, split-phase induction motor*

Types of single-phase induction motor

Figure 6.24   *Split-phase motor*

Split-phase motor

The split-phase motor has two stator windings and they are
1) a main winding which is termed the "run" winding
2) an auxiliary winding called the "start" winding

Windings connections

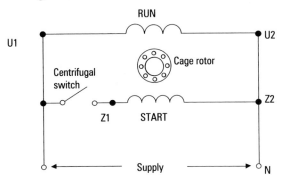

Figure 6.25

The run winding is wound with heavier gauge wire than the start winding and has less resistance than the start winding which is wound with very fine gauge wire. The inductive reactance of the run winding is however greater than that of the start winding.

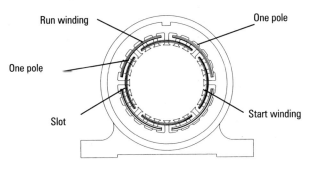

Figure 6.26   *Showing the two windings of a split-phase motor*
             Each winding has 4 poles (2 pairs).

As the inductive reactance of the run winding is greater than that of the start winding the current in the two windings will be out of phase with each other and reach maximum values at different times during each cycle. Hence the magnetic flux produced by these currents will also reach maximum and minimum values at different instants of each cycle.

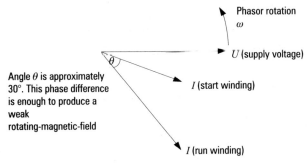

Figure 6.27

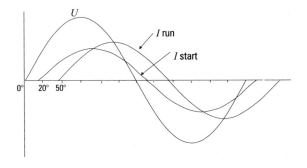

Figure 6.28

It can be seen that the current in the run winding lags further behind the supply voltage than that in the start winding because of its higher reactance. Consequently the current in the start winding reaches a maximum value before that in the run winding.

The start winding is short-time-rated and will burn out if left in circuit. To prevent this happening it is automatically cut out by a centrifugal switch when the motor reaches about 75% full speed. The moving arms for the centrifugal switch are fitted on the cage-rotor shaft (Figure 6.29) and operate contacts fitted on the inside of the stator frame.

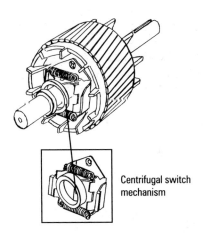

Figure 6.29   *Cage rotor*

Changing the direction of rotation
To change direction of rotation reverse the connection of one winding only i.e. start or run.

Typical starting torque
– 175% to 200% of full load torque

Starting current
– approximately 600% to 900% of full load current

Applications:
Domestic appliances, for example:
- washing machine agitators
- tumbler dryers

Capacitor start-induction run motor

Figure 6.30   Capacitor start-induction run motor

The starting characteristics of a split-phase motor can be improved by connecting a capacitor in series with the start winding. This type of motor is called a CAPACITOR START-INDUCTION RUN motor.

Windings connections

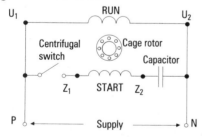

Figure 6.31

The current through the run winding lags the supply voltage due to the high inductive reactance of this winding and the current through the start winding leads the supply voltage due to the capacitive reactance of the capacitor. The phase displacement (angle θ) in the currents of the two windings at starting is now approximately 90° , (Figure 6.32) therefore the

magnetic flux set up by the two windings is much greater at starting than in the split-phase type motor, and this produces a relatively higher starting torque.

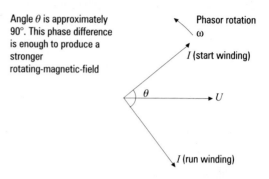

Angle θ is approximately 90°. This phase difference is enough to produce a stronger rotating-magnetic-field

Figure 6.32

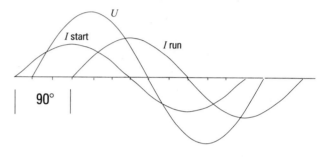

Figure 6.33

Typical starting torque
– 300% of full load torque

Typical starting current
– 500% of full load current

Note: The centrifugal switch operates the same as with the split-phase motor but now it disconnects the start winding and its series connected capacitor.

Application:
• Compressors
• Refrigerators
• Other applications involving a hard starting load
Typical motor ratings are 100 W to 500 W. However, larger sizes are available.

Changing direction of rotation
To change direction of rotation – reverse start or run winding connections – NOT both

Capacitor start – capacitor run motor

This type of motor is designed to operate with the start winding and its series capacitor permanently connected to the supply. It has two capacitors connected in parallel in the start winding circuit for starting purposes (Figure 6.34).

Figure 6.34   Capacitor start – capacitor run motor

Windings connections

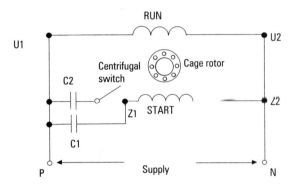

Figure 6.35   C1 is a paper or polypropylene capacitor
C2 is a large electrolytic capacitor

The centrifugal switch cuts out C2 and the smaller capacitor C1 is left in series with the start winding being continuously rated.

The starting torque and current is similar to the induction run type motor.

Advantages of capacitor start – capacitor run motors

Two main advantages with this type of motor are:
1)   Better running torque
2)   Improved power factor

Changing direction of rotation:

To change direction of rotation reverse start or run winding connections – NOT both.

Remember
The capacitor start capacitor run motor has two capacitors.
Two main advantages are improved power factor and running torque.
The centrifugal switch cuts out the large capacitor only.

Shaded pole motor

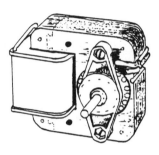

Figure 6.36   Shaded pole motor

The shaded pole motor is a very small fractional horsepower motor of the cage rotor type. It works on the principle of the movement of magnetic flux across the pole faces of the stator, cutting the rotor bars on the cage rotor, for starting.

Basic construction

The stator has two salient poles and each pole has a slot cut away to split the pole faces into two portions. One portion of each pole face is fitted with a copper shading ring. These heavy copper rings are magnetically displaced to create an artificial phase shift. A single coil is wound onto a bobbin which is fitted to the laminated core. Figure 6.37

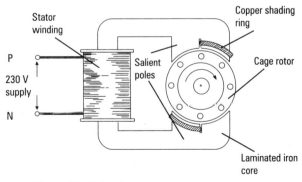

Figure 6.37   Shaded pole motor

Direction of rotation is from the unshaded to the shaded part of the pole face.

Operation of simple shaded pole motor

When the single-phase supply is connected to the stator winding it produces a pulsating magnetic flux in the pole pieces. Most of this flux bypasses the shading rings and crosses the air gap to the rotor but some of the flux cuts through the shading rings and induces an e.m.f. and a current in them. This induced current sets up a magnetic flux around the shading rings which opposes the main field flux when the current grows, and aids it when the current decays. Figure 6.38

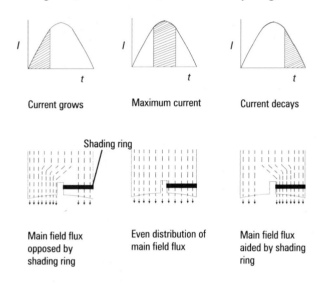

Current grows   Maximum current   Current decays

Main field flux opposed by shading ring   Even distribution of main field flux   Main field flux aided by shading ring

Figure 6.38   Shaded pole motor
Movement of flux across pole faces
(over one half cycle)

Consequently both portions of the pole face reach maximum and minimum values of flux at different instants of each half cycle and so a movement of flux occurs across each pole face. This movement is just sufficient to cut the rotor bars and start the motor.

Typical applications:

- Small fans, for example oven fans
- Hair-dryers
- Record player turntables and tape decks
- Office equipment
- Display drives

Direction of rotation

The direction of rotation depends on whether the shading rings are fitted on the right or left side of the pole pieces so this type of motor is seldom reversible but some are made to allow the shading ring to be fitted to either portion of the split pole. Provided both rings are changed over the motor will turn in the opposite direction.

This type of motor is usually the cheapest form of induction motor.

Remember
The shaded pole motor is only a very small fractional horsepower motor.

It has two "heavy" copper rings called shading rings on each salient pole piece to produce a movement of flux across them.

The direction of rotation is from the unshaded to the shaded portion of the pole face.

Universal (a.c./d.c.) motor

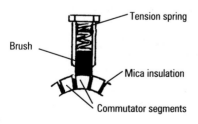

Figure 6.39   Universal motor

The universal motor can be operated on either an alternating or direct current supply and is very similar in construction to the d.c. series wound motor. Figures 6.40, 41 and 42.

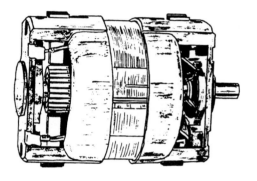

Brush holder

Field pole
Field windings

Figure 6.40   Field system and end plates

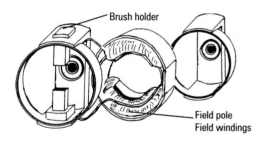

Tension spring

Brush

Mica insulation

Commutator segments

Figure 6.41   Brush holder

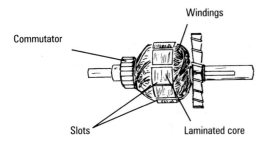

Figure 6.42   *Armature*

The main difference being that the yoke is made up of laminations to reduce eddy currents and prevent overheating when operated from an a.c. supply.

The armature and the field windings are connected in series exactly like the d.c. series motor as shown in Figure 6.43.

Windings connections

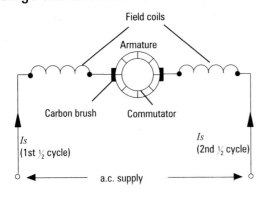

Figure 6.43

Operation

The basic principle of operation is due to interaction between the main field and the armature field which is produced by the same current (the supply current Is).

First half cycle

I_S flows through the left hand side field coil, then it passes into the armature conductors via the commutator and then through the right hand side coil. Figure 6.43

Second half cycle

I_S flows in the opposite direction.

Because the supply current is "alternating", both the main field polarity and armature conductor polarity will change at the same moment in time, therefore the motor will continue to run in the same direction. These changes can be seen more clearly by applying "Flemings Left Hand Rule" to Figures 6.44 and 45.

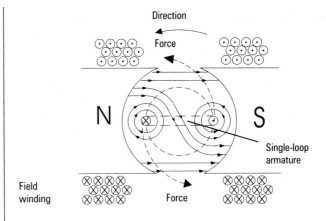

Figure 6.44   *Field interaction first half cycle*

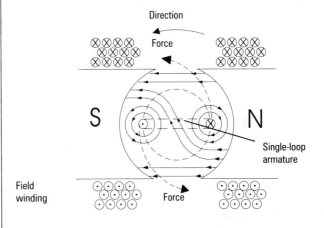

Figure 6.45   *Field interaction second half cycle*

Advantages
1)   Can be used on a.c. or d.c.
2)   Higher running speeds than the "induction type" motor (typical – 8 000 to 12 000 rev/min maximum)
3)   Cheap to produce
4)   High starting torque
5)   The higher running speed plus a good power factor develops more output power for a given size than any other single phase motor.

Disadvantages
1)   Commutation on a.c. is not as good as on d.c.
2)   Brush and commutator wear is more rapid when used on a.c., due to additional sparking at the brushes caused by the effect of inductance.
3)   Brushes are also prone to wear, especially when the motor is running at high speeds for long periods.
4)   If the load on the motor is reduced the speed will rise rapidly.

Application:

Universal motors are widely used in domestic appliances such as:

- Vacuum cleaners
- Food mixers
- Hair dryers
- Washing machines – spin dryers
- Portable tools, for example drills, saws etc.

Note: These motors do not usually exceed ¼ or ⅓ horsepower.

Changing the direction of rotation

To reverse direction of rotation change either the field coil connections or armature connections – NOT both.

Remember

Universal motors operate on a.c. or d.c.

The armature and field are connected in series, like a d.c. series motor.

Change armature or field connections to reverse direction of rotation.

This motor can run at very high speeds.

Points to remember ◀ — — — — — — — —

Single-phase motors

The split-phase induction motor has _____ windings, a _____ winding and a _____ winding connected in _____ across the supply.

On a split-phase motor the _____ cuts out the start winding to prevent it from _____.

To change the direction of rotation of a capacitor-start induction-run motor reverse either the _____ or _____ winding connections.

The shaded pole motor has two salient _____ and each pole face is split into _____ portions. One portion of each pole face is fitted with a copper _____ to produce a movement of _____ across them.

Universal motors can operate on either _____ or _____ supplies, and can run at _____ speeds.

Part 4

The control of a.c. motors

A.C. motor starters

Motor starters must all be able to:

- connect and disconnect the motor to the supply in a safe manner
- give the motor protection from abnormal overloading

Motor starters should be able to:

- prevent the motor from restarting after a supply failure or severe undervoltage (necessary, except for a few special applications)
- limit the current taken by the motor when starting (necessary for larger motors)

Where applicable motor starters should also:

- control the starting torque
- reverse the direction of rotation of the motor.
- control the speed of the motor

Important Note

Where applicable, the following wiring regulations to BS 7671 should be taken into consideration when designing motor control circuits:

> 130-06-01
> 435-01-01
> 462-01-01
> 476-02-03
> 525-01-02
> 537-04-01 and
> 552-01-01 to 552-01-05

Types of three-phase motor starter

There are several types of starter used for starting three-phase "cage rotor" induction motors including

- Direct-on-line
- Star-delta
- Auto transformer

Direct-on-line (D.O.L) starter

Direct-on-line starters are used for starting the majority of small three-phase cage rotor induction motors.

With this type of starter the stator windings of the motor are connected directly to the main supply lines L_1, L_2 and L_3. Therefore there are 400 volts across each stator winding on starting. Since the motor is at rest when the supply is switched on, the initial starting current is heavy and may cause some disturbance to the electricity supply (for example lights could flicker or dim). To overcome these problems supply companies limit the use of direct-on-line starting of induction motors above certain power ratings, typically 7.5kW. Where doubt exists a check should be carried out to ensure direct-on-line starting is permissible.

The initial current surge on starting can be six to ten times the full load current and the initial starting torque is about 150% of full load torque.

Application

Direct-on-line starters are suitable for "light" starting loads.

Remember

All motor starters consist of a mains circuit, which switches the windings of a motor, and an auxiliary circuit which is used to control the switching functions.

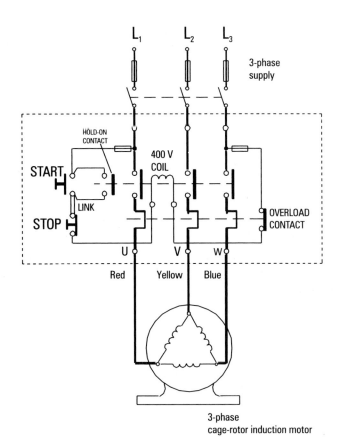

Figure 6.46   Direct-on-line starter

Power (main) circuit shown by thick lines.
Control (auxiliary) circuits shown by thinner lines.
The fuses provide short-circuit protection.
The overloads may be thermally or magnetically operated.

Operation of direct-on-line starter (Figure 6.46)

Press the start button and the contactor coil energises. The main contacts and "hold-on" contact close and the motor starts to run.

When the start button is released the contactor coil remains energised via the "hold-on" contact and the motor continues running.

Press the stop button the contactor coil de-energises, the main contacts open and the motor stops.

Overload protection

If the motor draws excessive current (overloads) the overload coils will trip out the overload contact, and this open circuits the contactor coil. The coil de-energises, the main contacts open and the motor stops. Should one phase fail the motor would be put under undue pressure to continue to work. This situation is called "single phasing" and the overload equipment should detect this and stop the motor.

All motors having a rating exceeding 0.37 kW must have control equipment with overcurrent protection.

No-volt protection

If there is a supply failure, the contactor coil will de-energise and the "hold-on" contact will open. Hence, the contactor coil cannot re-energise again (when the supply is reinstated) until the start button is pressed by the operator.

Undervoltage protection

If the supply voltage falls to about 80% of its nominal value the contactor coil de-energises and trips out the contactor. The operator will have to press the start button to re-energise the contactor because the "hold-on" contact will be open again.

It is most important that No-volt/Undervoltage protection is provided to prevent automatic starting of machinery after a failure of the supply. For example a machinist could be in serious danger if the lathe restarted unexpectedly when the supply was re-instated.

Choosing a direct-on-line starter

When choosing a direct-on-line starter for a particular application ensure that:

- the enclosure is suitable for the environmental conditions. General types available are:

    steel box/polycarbonate cover,
    all polycarbonate
    steel box and cover,
    heavy cast iron

- the rated operating voltage and current of the contacts are suitable

- it can handle the motor rating (kW)
- the voltage of the operating coil is suitable
- the overload device can handle the full-load-current of the motor
- the push-button arrangement is suitable
- the contactor has the correct number of normally closed and normally open auxiliary contacts to meet control circuit requirements.

Remote start/stop control of a direct-on-line starter

When connections are required for remote start/stop buttons it should be remembered that:

- extra start buttons are wired in parallel with existing ones
- extra stop buttons are wired in series with existing ones

To aid understanding of the control circuit in Figure 6.46 a schematic (straight line) diagram is used, as shown in Figure 6.47.

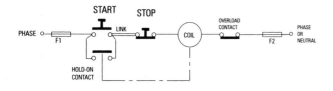

Figure 6.47   Control circuit for direct-on-line starter

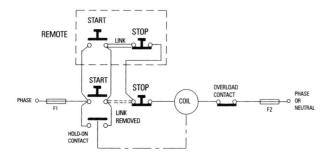

Figure 6.48   Control circuit with remote start/stop

Note: When the control circuit is connected between the phase and neutral, fuse F2 can be omitted or replaced by a link in Figures 6.47 and 6.48

Notice that only three wires are required to connect in the remote start/stop control.

To add on another remote start/stop position simply remove the link, run three more wires out and connect as above.

Forward and reverse direct-on-line starter

As we have seen, to reverse a three-phase motor any 2 phases have to be changed over. The forward and reverse starter is designed to do this automatically. This consists of 2 contactors, one with the supply connected directly in the sequence $L_1 L_2 L_3$ to UVW, the other connecting $L_1 L_2 L_3$ to WVU respectively. The circuit diagram is shown in Figure 6.49 opposite.

Forward operation

Press the forward button.
The forward contactor coil energises and the forward contactor closes, the electrical interlock F.C.1 for the forward contactor opens. This is to stop the reverse contactor coil becoming energised at the same time as the forward contactor coil.

The hold-on contact for the forward F.C.2 closes and the motor starts to run.

Release the forward button.
The forward coil remains energised via hold-on contact F.C.2.

The motor continues to run.

Press the stop button.
The forward contact coil de-energises, the forward contactor opens and the motor stops.

A similar operation occurs when the reverse button is pressed, but with the other contactor.

Try this
Explain the reverse operation of the direct-on-line starter.

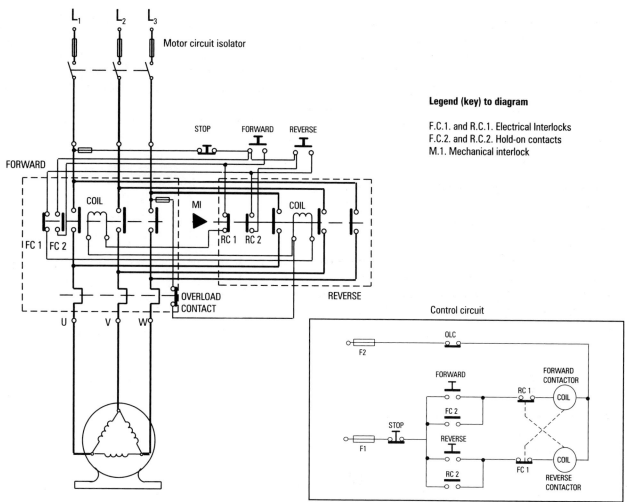

Legend (key) to diagram

F.C.1. and R.C.1. Electrical Interlocks
F.C.2. and R.C.2. Hold-on contacts
M.1. Mechanical interlock

Control circuit

Figure 6.49   *Forward and reverse direct-on-line starter*

Points to remember ◀ – – – – – – – – – – – –

Three-phase motor starters

Direct-on-line starters are suitable for _____ starting loads.

This type of starter has three _____, one in each phase, to protect the motor when it starts drawing _____ current from _____ .

No-volt protection is provided to prevent _____ _____ of machinery after a supply _____ .

Start buttons are wired in _____ and stop buttons are wired in _____ when connecting _____ stop/start controls.

Forward/reverse D-O-L starters have _____ interlocks and a _____ interlock to prevent both _____ being energised at the same time.

Remember

The electrical interlocks and the mechanical interlock prevent both contactors being energised together.

Two phases are changed over to reverse the direction of the motor.

Part 5

Reduced voltage starting methods

Star-delta starting
This is the most common method of reduced voltage starting.

On starting, the stator windings are first connected in a "star" configuration. This is to reduce the voltage across each winding.

i.e.

$$U \text{ phase} = \frac{U_L}{\sqrt{3}} = \frac{400}{1.732}$$

$$= 230 \text{ V approximately}$$

230 V is approximately 58% of 400 V.

The motor is then run up to speed with 230 V across each winding and when it has attained approximately full speed the starter is switched to connect the windings in "delta" configuration with the full line voltage of 400 V across each winding.

Sufficient time must be allowed for the motor to run up to speed before switching from star to delta to prevent the possibility of heavy overloads and damage to the motor when using this method.

Typical values for starting a three-phase cage rotor induction motor by star-delta are:
- initial starting current 2 to 4 times full-load-current
- initial starting torque 50% of full-load-torque

The starting current and starting torque is reduced to one third of that which would occur if the motor was started direct-on-line.

Application
Used for starting on "no-load" or "light load".

Motors of all sizes up to about 13 kW can be star delta started.

Manual star delta starter operation

This method of star delta starting requires the operator to manually switch the starter into a star position and then when the motor runs up to speed change over to delta.

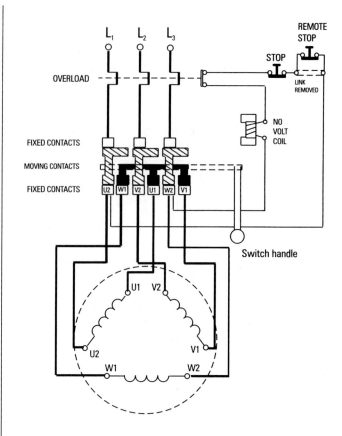

Figure 6.50   Starter switched to star

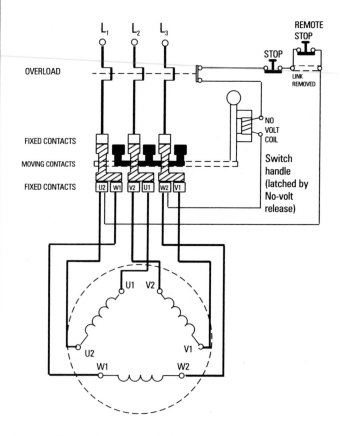

Figure 6.51   Starter switched to delta

Operation

Move the change-over switch downwards to the "star" starting position, Figure 6.50. (Supply lines L_1, L_2 and L_3 are connected to U_2, V_2 and W_2 respectively, and U_1, V_1 and W_1 are connected to star.)

Allow sufficient time for motor to run up to speed.

Move the change-over switch upwards to the "delta" running position, Figure 6.51. (L_1 is connected to U_2 and W_1, L_2 is connected to V_2 and U_1, L_3 is connected to W_2 and V_1) and leave the switch in this position.

The no-volt coil energises and holds the change-over switch in the "delta" position by an electro-magnetic mechanism. The motor continues running.

Press the stop button, the no-volt coil de-energises, the switch mechanism releases, the switch contacts open and the motor stops.

The no-volt coil de-energises also when:
- an overload situation occurs
- there is a supply failure
- the supply voltage falls below a certain value (undervoltage situation)

Consequently the motor stops and cannot restart until the change-over switch is moved downwards again to the star starting position by the operator.

Auto star delta starter

Unlike the manually operated starter the auto star delta starter only requires a push of the start button to set the process in motion.

Operation

Press the start button and the main contactor coil energises.

The star contactor coil and timer (T) both energise via the timer contact (Y) and "delta" interlock contact.

The main contacts close and connect the three-phase supply lines to the motor winding terminals U_1, V_1 and W_1.

The star contacts close and connect motor winding terminals, U_2, V_2 and W_2 together to form the "star point".

The motor starts to run on reduced voltage with 230 V across each winding.

Release the start button and the main contactor coil remains energised via the hold-on contact.

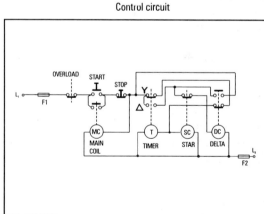

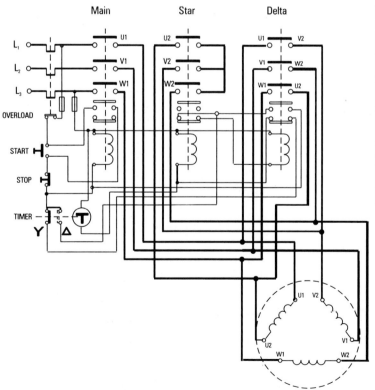

Figure 6.52   Auto star delta starter

The star contactor remains energised via the hold-on contact and the timer contact (Y).The motor continues to run on reduced voltage.

The timer "times-out" and switches the change-over contact from the Y position to the Δ position.

The star contactor coil de-energises and the delta contactor coil energises via the timer contact Δ and the star interlock contact.

The delta contacts close and connect the following motor winding terminals together U_1 to V_2, V_1 to W_2 and W_1 to U_2 to form the delta connection of the windings.

The motor continues to run on full supply voltage with 400 V across each winding.

The timer de-energises (resets) to the star position, and the delta contactor coil remains energised via its own hold-on contact.

Press the stop button and the main and delta contactor coils de-energise, the main contacts open and the motor stops.

Overload protection

If the motor draws excessive current the overload coils will trip out the overload contact to de-energise the contactor coils.

No-volt and undervoltage protection

If either the supply fails or the supply voltage falls drastically the contactor coils will de-energise and they cannot re-energise until the operator has pressed the start button because the "hold-on" contact opens the circuit to the coils when the main contactor de-energises.

Auto-transformer starter

The auto-transformer is used where only three leads are available at the motor terminal box or because star delta starting torque is insufficient.

Several transformer tappings are provided to reduce the voltage across the stator windings on starting but only one of these tappings is selected as the starting tapping.

An example of tapping arrangements is:

50, 60 and 75% of line voltage

Other tappings can be provided but it is not practical to go below 50% as the torque produced is insufficient.

The correct tapping for an auto-transformer starter is the lowest one at which the motor begins to turn and gain speed.

Application

Auto-transformer starters are used for motors of medium and large size (up to about 75 kW) on light starting loads.

Examples are:

    centrifugal pumps
    fans
    compressors
    mills

Remember
Auto transformer starters are employed when star delta starting torque is insufficient.

Hand operated auto transformer starter

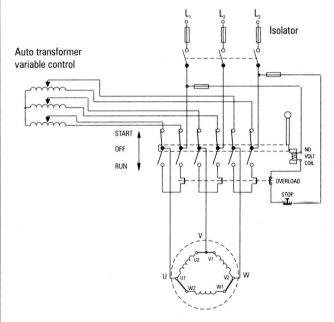

Figure 6.53

Operation

Close the main isolator

Move the change-over switch to start position and reduced voltage from the auto transformer is applied to the stator.

Move the change-over switch to the run position *only when sufficient speed has been reached.* Full line voltage is now applied to the stator. The change-over switch is held into the run position by an electromagnetic latch on the no-volt release coil.

The change-over can be carried out automatically using a timing relay.

Starting three-phase motors on load

The starters previously mentioned were used to start 3-phase cage rotor induction motors with "no-load" or "light loads". When it is required to start a motor against a "heavy load", a 3-phase slip-ring motor with a wound rotor can be used. This is started by a rotor resistance type starter. (Figure 6.54)

Wound rotor motor starters

The starting torque of this type of motor depends upon the total resistance in the rotor circuit, therefore the external resistances can be arranged to give maximum motor torque on starting from standstill. However, the starting torque is usually kept down to about 150% of full-load-torque, with a starting current of about 150% of full-load-current. This is done to reduce the disturbance to the mains supply and the "shock" to the driven machine.

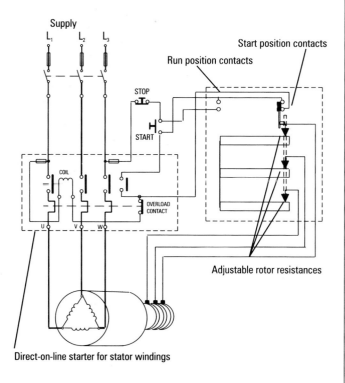

Direct-on-line starter for stator windings

Figure 6.54   Hand operated rotor resistance starter

Rotor resistance starter operation

Press the start button, the contactor coil energises pulling the main contacts and hold-on contacts closed. This connects the three-phase supply to the motor stator terminals U, V and W. The motor starts to run with maximum resistance in the rotor circuit.

Keep the start button depressed until all resistance has been taken out of the rotor circuit. This ensures that the motor is not accidentally left running with some of the rotor resistance still in circuit. When all the

resistance is taken out the "Run" contact closes and short circuits the start button. Resistances are now "short circuited" in this position.

Release the start button, the contactor coil remains energised via the "run" and "hold-on" contacts and the motor continues running.

Press the stop button, the contactor coil de-energises and the motor stops.

The motor cannot be restarted until the resistance controller is returned to the start position.

If the main supply fails the coil on the stator starter is de-energised and the motor stops.

Remember
Rotor resistance starters are used for starting three-phase slip-ring motors with wound rotors.

Points to remember ◀ — — — — — — — — — — — —

Reduced voltage starting methods

The most common method of reduced voltage starting is _____ starting.

With this starting method, the stator windings are connected in _____ to reduce the _____ across each _____ on starting, and then when the motor has attained approximately _____ speed the starter is switched to connect the windings in _____ configuration with the full _____ voltage of _____ across each _____.

When star/delta starting a three-phase _____ rotor induction motor the initial starting current is typically 2 to 4 times _____ and the initial starting torque is typically 50% of _____.

Auto transformer starters are used when star/delta _____ _____ is insufficient.

This type of starter has several transformers _____ to reduce the voltage across the _____ windings on starting.

Rotor resistance starters are used for starting _____ _____.

Part 6

Types of single-phase motor starter

A simple manually operated starter which could be used to start a split-phase or a capacitor start motor is shown below. (Figure 6.55)

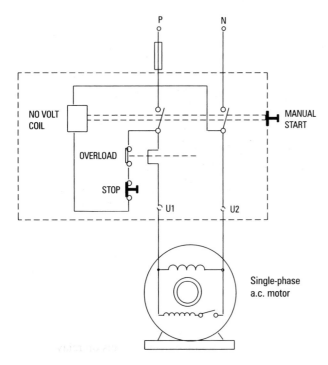

Figure 6.55

Operation

> **Press the start button**, the double-pole switch closes, the supply is connected to the motor and the motor starts. At the same time the No-volt coil energises.

> **Release the start button**, the double-pole switch is held closed by the No-volt latching mechanism and the motor continues running.

> **Press the stop button**, the No-volt coil de-energises, the latching mechanism releases, the double-pole switch opens and the motor stops.

Single-phase direct-on-line starter circuit

The operation of this starter is the same as for the 3-phase direct-on-line starter in Figure 6.46, since the control circuit is the same.

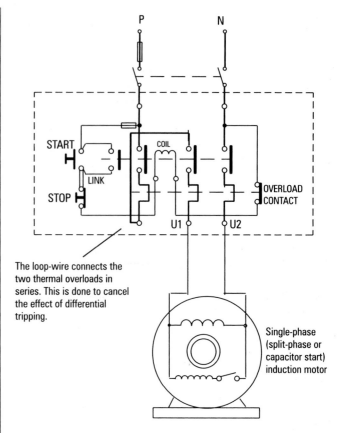

The loop-wire connects the two thermal overloads in series. This is done to cancel the effect of differential tripping.

Figure 6.56

Starting a universal (series wound) motor

A universal motor is usually started direct-on-line, as shown in Figure 6.57. A change-over switch can be used to reverse the direction of rotation (Figures 6.58 and 6.59).

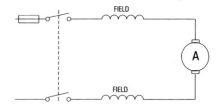

Figure 6.57   Double pole switch for starting

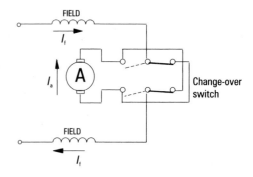

Figure 6.58   Reversing using a change-over switch

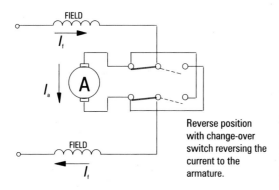

Figure 6.59

The current through the armature I_a is reversed and the current through the field coils I_f remain the same.

This is often used on electric drills so that the chuck can be reversed.

This type of switching arrangement can also be used to reverse the direction of rotation of a split-phase or a capacitor-start type single-phase motor.

Motor temperature protection

Positive temperature co-efficient type thermistors are widely used to prevent damage to motor windings from overheating since their resistance changes very rapidly at a critical temperature (see graph). It is this feature of a sudden change in resistance which is used to initiate the switching off of a motor.

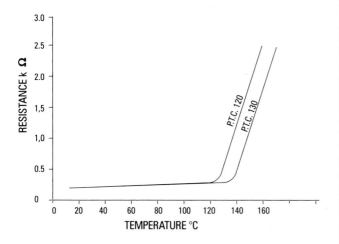

Figure 6.60   *Typical characteristics for "barium titanite" P.T.C. thermistors.*
Resistance is low and relatively constant at low temperatures. The rate of increase becomes very rapid at the SWITCHING TEMPERATURE POINT. Above this point the characteristic becomes very steep and attains a high resistance value.

A normal three-phase motor has three thermistors connected in series embedded in each phase winding of the stator.

As they carry only very small currents (mA) thermistors are used in conjunction with an amplifier or relay unit for the control of a motor (Figure 6.61).

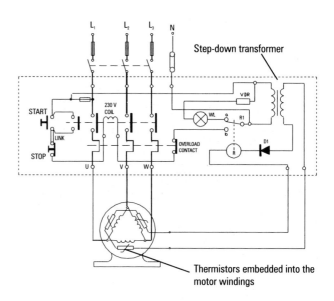

Figure 6.61   *A three-phase direct-on-line starter incorporating thermistor protection*

Key:

D_1	Diode
R	DC relay coil
R_1	Change-over contacts on relay
WL	Warning lamp
VDR	Voltage dropping resistor

Operation of thermistors

Under normal conditions (i.e. safe temperature) sufficient current flows in the thermistor circuit to energise the relay (R) therefore the change-over contact (R_1) is held in position (b) and the contactor can operate as normal.

Should the motor winding temperature reach the "switching temperature point" the rapid increase in resistance will reduce the current, the relay will de-energise and the change-over contact will move to position (a) so the contactor coil will de-energise and the warning lamp indicate a fault condition.

When the windings and thermistors have cooled down a few degrees the relay will energise again and the lamp will go out. The contactor can be energised again by pressing the start button.

The choice of thermistor, for motor protection, depends upon the switching temperature point and the class of motor insulation (i.e. switching point of 120° C for Class E and 130° C for Class B).

Motor enclosures

Once the most suitable type of motor for a particular job has been selected it is necessary to ensure that the motor enclosure is also suitable for its working environment. Various types are shown below.

Figure 6.62   Totally enclosed fan ventilated motor (TEFV)

Figure 6.63   Drip proof motor

Figure 6.64   Flameproof motor
BASEEFA Mark
Where equipment is to be installed in potentially hazardous areas it should carry the appropriate symbol.

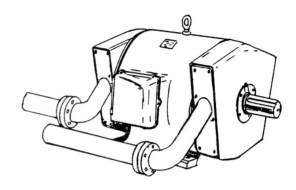

Figure 6.65   Pipe ventilated motor

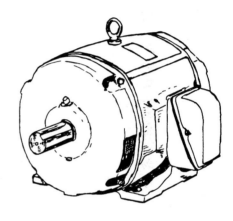

Figure 6.66   Screen protected motor

Typical applications

Totally enclosed	boilerhouses, steelworks, outdoor winches
Screen protected	general purpose, engineering workshops
Drip proof	pump rooms, laundries
Pipe ventilated	cement works, flour mills, paper mills
Flameproof	chemical works, gasworks, oil plants

Remember
When choosing a motor for a particular application ensure that its enclosure is suitable for the environmental conditions.

Isolation, switching, control and protection of motors

For safe and efficient installation and use of motors a sequence of control must be followed. An example of this is shown in Figure 6.67.

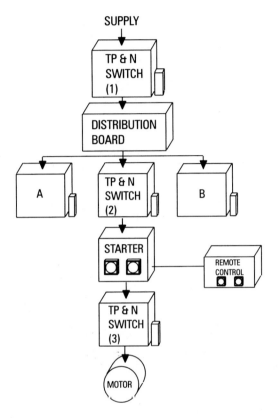

Figure 6.67

Function

The TP &N switch (1) isolates the electrical supply to the "group" of motor circuits.

The distribution board provides short-circuit and overload protection for the motor circuits. A and B go to other motors with similar starter and switches.

The TP &N switches(2) isolate the motor circuits so that work can be carried out on the starters and control devices.

The starter offers control of the motor:

>**functional switching** – to switch off the motor when not in use;
>**emergency switching** press the stop buttons to isolate the motor in an emergency;
>**no-volt protection** – to disconnect in the event of mains failure;
>**motor overload protection** – should the motor become overloaded.

The TP & N switch (3) is for local isolation so that mechanical maintenance can be carried out.

Try this
For safety reasons some of the devices shown in Figure 6.67 should have facilities for locking them in the OFF position. Find out which these are, list them below and give reasons why they should need to be locked.

Suitable ways of electrically connecting a motor to a local isolator or starter.

As motors vibrate when they are working they require a flexible method of wiring to them.

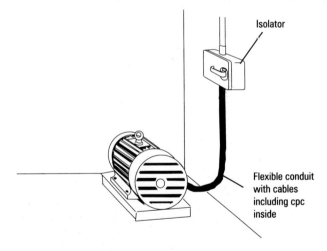

Isolator

Flexible conduit with cables including cpc inside

Figure 6.68   Motor connected using flexible conduit

In most cases a flexible conduit is used with single insulated conductors installed inside. The flexible conduit may be metal, fibre or PVC but in any case a separate circuit protective conductor must be installed.

Remember
A separate cpc must be run in flexible conduit to ensure the motor is safely connected to earth.

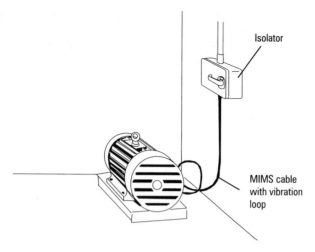

Figure 6.69   Motor supplied with MIMS cable

An alternative to flexible conduit is mineral insulated metal sheathed cable. This will have to be selected with the appropriate number of cores for the type of motor. A loop is usually left in the cable to absorb the vibration. The copper sheath of the cable acts as the circuit protective conductor.

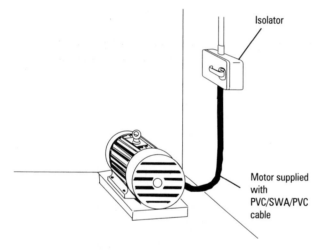

Figure 6.70   Motor supplied with PVC/SWA/PVC cable

PVC insulated steel wire armoured cable can be used to connect the motors to their supply. In this case care must be taken not to put the terminations under undue mechanical stress.

Points to remember ◄ – – – – – – – – – –

Single phase motor starters/thermistor protection/motor enclosures/installation of motors

Split-phase or capacitor-start single-phase induction motors could be started by a _____ operated type starter or a _____ type starter.

Universal motors are usually started _____ and a _____ switch can be used to_____ the direction of rotation.

P.T.C. type_____ are widely used to protect motor windings from _____ .

If the motor winding temperature _____ , and reaches the switching temperature point of the thermistor, its resistance_____ rapidly, thus the current in the control circuit _____ and the contactor coil _____ .

The most suitable type of motor enclosure for a cement works would be _____ and for a chemical works _____ .

Emergency stop buttons are connected in _____ in a single-phase or three-phase motor control circuit to enable the motors to be stopped in an _____ situation.

Three suitable ways of electrically connecting a motor to absorb vibration are:

1. _____

2. _____

3. _____

Self assessment short answer questions

1. Complete the diagram by including
   (a) the field around the conductor
   (b) the stator field. Indicate the direction of the force applied to the conductor and the direction of rotation of the rotor.

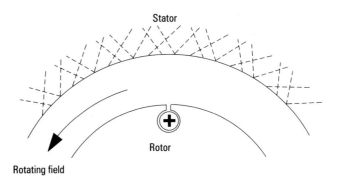

Figure 6.71

2. Calculate:
   (a) the synchronous speed of a twelve pole a.c. motor when connected to a 50 Hz supply,
   (b) the rotor speed if the percentage slip is 4%.

3. (a) Draw a labelled circuit diagram of a single-phase capacitor-start induction run motor.
   (b) State the purpose of the centrifugal switch fitted to this type of motor.

4. (a) Sketch a simple diagram of a cage rotor and label it.
   (b) State why the cage rotor core is laminated.

5. Describe briefly the procedure for reversing the direction of rotation for EACH of the following a.c. motors.
   (a) single-phase capacitor-start induction-run
   (b) three-phase cage-rotor induction
   (c) three-phase wound-rotor induction

7. The diagram below shows the control circuit for a direct-on-line starter.
   Redraw the diagram and include a remote stop/start control.

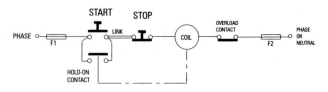

Figure 6.73

6. In Figure 6.77

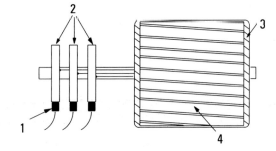

Figure 6.72

(a) Identify (i) the type of rotor shown
    (ii) the component parts numbered 1, 2, 3 and 4.
(b) State the function of the components numbered 2.

8. P.T.C. type thermistors are used to prevent motor windings from overheating.
   (a) State where the thermistors are fitted on a three-phase induction motor.
   (b) Describe briefly how this type of thermistor protects the motor windings from overheating when used in conjunction with a D-O-L starter.

7

Transformers

Answer the following questions to remind yourself of what was covered in Chapter 6.

1. State three types of starter used for 3-phase cage-rotor induction motors.

2.
   (a) What is the purpose of the shading rings on a shaded pole motor?
   (b) In which direction does the rotor of this type of motor rotate.
   (c) Give two applications for this type of motor.

3. A four pole induction motor runs at 23.5 rev/s when connected to a 50 Hz a.c. supply. Calculate the percentage slip of the motor.

4. State the purpose of the following devices in motor circuits:
   (a) thermal overloads
   (b) no-volt coils
   (c) thermistors

Part 1

Purpose of a transformer

The transformer is an extremely useful piece of electrical equipment. Its main use is to take an a.c. supply at one voltage and produce from this another a.c. supply. The voltage of the second supply may be quite different from that of the first and the process can isolate one from the other, a feature which makes the transformer ideal for the provision of safety services.

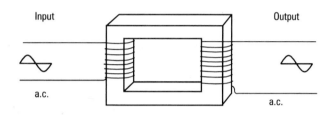

Figure 7.1   Double-wound transformer

Construction and enclosures

Small power transformers for use in equipment are usually of open construction, with varnished windings, and are air cooled.

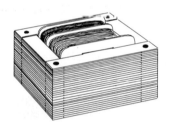

Figure 7.2   Small air-cooled power transformer

Larger transformers for medium power applications may be housed in metal tanks which are then filled with mineral oil which helps to insulate the windings. Excess heat is dissipated through the sides of the tank.

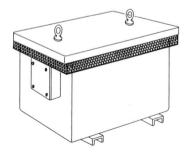

Figure 7.3   Oil -cooled transformer

Large power transformers are enclosed in a metal tank with cooling tubes fitted to the outside through which the oil is free to circulate. The oil serves as an insulating and a cooling medium as the natural convection of the liquid carries away the heat from the windings as well as insulating one from the other.

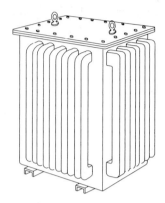

Figure 7.4   Large oil filled transformer

Any large item of oil-filled plant such as a transformer or circuit breaker must be mounted over a pit or surrounded by a low wall which will contain all the oil in the event of leakage.

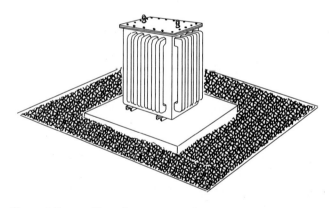

Figure 7.5   Transformer mounted over a pit

In the case of the pit, this is usually filled with loose stones or gravel to prevent anyone from falling in but will still hold all the oil contained in the tank or tanks if a leak should happen.

Transformer core arrangements

The core is built up around the coils using laminations which have been stamped out to give the correct shape when fully assembled.

Small power transformers are usually of the "core" or "shell" type assembly.

A core type will usually have its windings evenly distributed over the two outer limbs of a rectangular core with a single "window" in the centre.

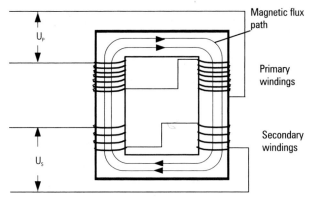

Figure 7.6   Core type transformer

The shell assembly uses a three limb construction with both windings mounted on the centre limb.

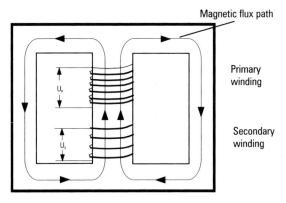

Figure 7.7   Shell type transformer

Leakage flux

Not all the flux produced by the primary winding will "link" with the secondary winding, this is termed the "leakage flux".

The" shell" type magnetic circuit has less leakage flux than the "core" type circuit (since both windings are wound on the centre limb there is a better magnetic linkage between them). Leakage flux tends to take place more at the corners of a magnetic circuit, this is why a ring-type (toroidal) core construction is often used for instrument and current transformers (Figure 7.8)

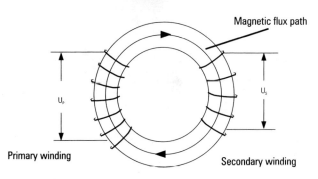

Figure 7.8   Ring type

If a transformer was wound as represented by Figure 7.9 with the primary and secondary windings on separate limbs of the core, the leakage flux will be quite a considerable amount due to the distance the windings are spaced apart from each other.

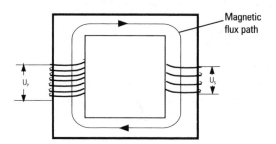

Figure 7.9

Operation of transformers

For all practical purposes, a transformer will consist of two coils of insulated wire wound around the same magnetic core.

One coil, which will be called the "primary" will be connected to an a.c. supply. The alternating current in the primary coil will set up an alternating magnetic flux in the core and this will link with the turns of the "secondary" winding. The alternating voltage thus produced will depend on the number of turns of wire in the secondary, just as the strength of the field will depend on the number of turns in the primary.

Assuming for the moment that there are to be no losses in the process then the Ampere-turns in the primary winding will be equal to the Ampere-turns in the secondary. Using the symbol N for the number of turns

$$N_p I_p \quad = N_s I_s \quad \text{(equation 1)}$$

On the same theme, and assuming that the device has no losses then we can say that;

$$\text{Volt Amps} \quad \text{in} \quad = \text{Volt Amps} \quad \text{out}$$

In other words
$$U_p I_p = U_s I_s \quad \text{(equation 2)}$$

Going back to equation 1, this can be re-written as
$$\frac{N_P}{N_S} = \frac{I_S}{I_P}$$

and similarly, Equation 2 can be written;

$$\frac{U_P}{U_S} = \frac{I_S}{I_P}$$

Which gives us;

$$\frac{U_P}{U_S} = \frac{N_P}{N_S} = \frac{I_S}{I_P}$$

This is often referred to as
THE BASIC TRANSFORMER EQUATION

Step down transformers

Transformers are widely used for electrical and electronic applications because they can change voltages from one level to another with relative ease.

If a transformer has 240 turns in the primary winding and 24 turns in the secondary it is said to have a **TRANSFORMER RATIO** of

$$10:1$$

In other words the primary voltage is reduced by a factor of 10.

This relationship can be expressed by the basic equation

$$\frac{N_P}{N_S} = \frac{U_P}{U_S}$$

Thus a transformer with a 10 to 1 ratio will have a secondary voltage which is relative to the primary namely;

$$U_S = U_P \times \frac{N_S}{N_P}$$

Taking the previous example;

If a transformer having 240 turns in the primary and 24 turns in the secondary is connected to a 125 V supply the secondary voltage will be;

$$U_S = U_P \times \frac{N_S}{N_P}$$

$$= 125 \times \frac{24}{240}$$

$$= 12.5 \text{ V}$$

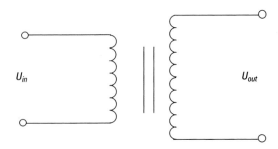

Figure 7.10   Step down transformer

Step up transformers

A transformer can just as easily be used for the purpose of raising the voltage and where higher voltages are required for a particular application the transformer provides a quick and easy solution.

For example if an item of equipment which is mainly operated at 24 V a.c. but for one particular process requires 600 V then the solution could be the inclusion of a transformer having a step up ratio of 25 to 1.

Given that the primary winding is to contain 50 turns then the secondary would need to be 25 times greater.

$$U_S = U_P \times \frac{N_S}{N_P}$$

$$= 24 \times \frac{1250}{50}$$

$$= 600 \text{ V}$$

Figure 7.11   Step up transformer

Example

A single-phase step-down transformer has 760 primary turns and 360 secondary turns.

Calculate the secondary
   (a) voltage if the primary voltage is 230 V a.c.
   (b) current if the primary current is 5 A

(a)
$$\frac{U_P}{U_S} = \frac{N_P}{N_S}$$

$$\therefore \frac{U_S}{U_P} = \frac{N_S}{N_P}$$

(turning both sides upside down makes it easier to transpose)

$$\therefore U_S = \frac{N_S}{N_P} \times U_P$$

$$= \frac{360}{760} \times 230$$

$$= 109V$$

(b)
$$\frac{U_P}{U_S} = \frac{I_S}{I_P} \qquad \therefore I_S = \frac{U_P}{U_S} \times I_P$$

$$= \frac{230}{109} \times 5$$

$$= 10.55A$$

Try this

A transformer is to be used to provide a 57.5 V output from a 230 V a.c. supply.

Calculate:
   (a) the turns ratio required
   (b) the number of primary turns, if the secondary is wound with 500 turns

---

Volts per turn

Since both windings of a double-wound transformer are "linked" by the same magnetic flux (Figure 7.6), the induced e.m.f. per turn will be the same for both windings. Therefore, the e.m.f. in both windings is proportional to the number of turns.

Thus:

| The volts per turn on primary winding | = | the volts per turn on the the secondary winding |

Or
$$\frac{U_P}{N_P} = \frac{U_S}{N_S}$$

Example

A double-wound 230 V/50 V single-phase transformer has 110 primary turns. Calculate the volts/turn on the primary and secondary windings.

$$\text{Volts/turn (primary)} = \frac{U_P}{N_P} = \frac{230}{110} = 2.09$$

Remember $\dfrac{U_P}{U_S} = \dfrac{N_P}{N_S}$ $\therefore N_S = \dfrac{U_S}{U_P} \times N_P$

$$= \frac{50}{230} \times 110$$

$$= 23.9 \text{ turns}$$

$$\text{Volts/turn (secondary)} = \frac{U_S}{N_S} = \frac{50}{23.9} = 2.09$$

Try this

A transformer with 500 primary turns and 125 secondary turns is fed from a 230 V a.c. supply. Calculate:
   (a) the secondary voltage
   (b) the volts per turn

Construction/operation of transformers

Small power transformers are _____ cooled.
Medium power transformers may be housed in
_____ _____ which are filled with
_____ _____ which helps to
_____ the windings.

Large power transformers are enclosed in a _____
_____ with _____ _____
fitted to the outside through which the oil _____
through.

Oil filled transformers are mounted over a _____
filled with loose _____ to contain all the
_____ in the event of a _____ .

The two basic core arrangements for single-phase
double-wound transformers are
1. _____, which has its windings evenly
   distributed over the two outer _____.
2. _____, which has both windings mounted on
   the _____ _____.
A current transformer has a _____ type core.

Operating principle – the _____ current in the
_____ winding sets up an alternating
_____ _____ in the _____
and this links with the turns of the _____
winding to induce an _____ into it by means of
_____ induction.

A step-down transformer has more _____ turns
and a step-up transformer has more _____ turns.

The basic transformer equation is:

$$\qquad = \qquad = \qquad$$

The volts per turn on the primary = the _____ on
the _____

Part 2

Isolating transformers

It is not always the case that transformers are used to change
the voltage and for reasons of safety it may be required to
provide a mains voltage supply which is not derived directly
from the mains supply.

This is the principle adopted in the BS3535 bathroom shaver
socket in which the 230 V shaver supply is provided by a socket
outlet which has no reference connection to earth potential. It is
therefore incapable of delivering an earth leakage current.

The same technique is used in the service and repair industries
where technicians are frequently required to work on live
electrical equipment.

If the equipment under repair is supplied through an isolating
transformer then the technician, although still exposed to
mains voltage, is not exposed to the dangers arising from
simultaneous contact with exposed or extraneous conductive
parts such as earthed apparatus or pipework.

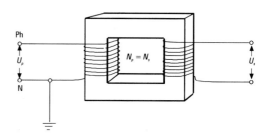

*Figure 7.12   Isolating transformer with no connection to
earth on the secondary winding.*

Voltage/current relationships

Because the transformer is a very efficient item of equipment
then for most practical purposes its efficiency can be assumed
to be very close to 100%.

That is to say, the power delivered to the primary winding is
assumed to be equal to the power delivered by the secondary.

i.e.

$$\text{Power in} = \text{Power out}$$

For example,
A 5 kVA transformer is supplied with 20 A at 250 V. Using this
very basic relationship we could deduce that such a device
would be capable of delivering 10 A at 500 V if the ratio
happened to be 1:2.

Alternatively, if the secondary voltage happened to be 100 V
the current would have to be 50 A in order to maintain the same
primary current and thus the ratio would be a step down in the
order of 2.5:1.

Auto-transformers

Not all transformers are of the double wound type (Figure 7.1). The auto transformer has only one winding and is capable of stepping up and down the voltage as effectively as the double wound variety.

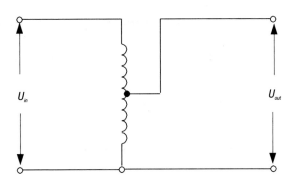

Figure 7.13   Step down auto-transformer

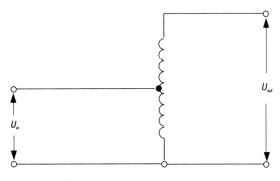

Figure 7.14   Step up auto-transformer

One important advantage of the autotransformer is that the part of the winding which is common to both primary and secondary current carries the difference between the two currents. Figure 7.15

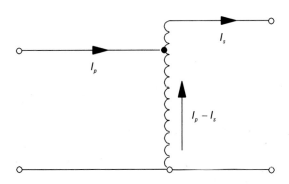

Figure 7.15

Where this type of transformer is used to make small adjustments in mains voltage the currents are very nearly equal and therefore the resultant current in the common part of the winding is quite small. For this reason, the common section can be wound in comparatively light wire resulting in considerable cost and weight savings as compared with double wound types.

The main problem with the auto transformer arises from the fact that it does not have an isolated secondary. Another unfortunate feature is that if the common terminal should become disconnected then full input voltage will appear at the output terminals.

Transformer windings

It is quite convenient to draw a transformer as two windings wound on separate limbs of a common core. Although this device would work it is not the most efficient form of construction. Better flux linkage is achieved by building up the windings in alternate layers or discs so that the flux produced by one winding is more readily adjacent to the turns of the other.

Concentric windings are used for all sizes of power transformers. This technique is used to build up a cylindrical coil assembly consisting of alternating layers of primary and secondary turns which are insulated from each other by interleaved insulation to prevent breakdown.

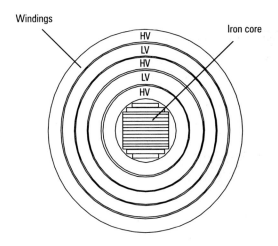

Figure 7.16   Concentric windings

Disc or sandwich winding involves assembling and interconnecting alternate sections of the winding by placing one on top of the other in a stacking arrangement.

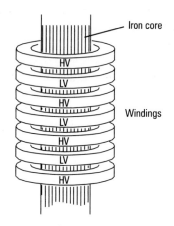

Figure 7.17   Disc or sandwich windings

Three phase transformers

Three phase transformers are an essential feature of transmission and distribution systems. They are to be found in power stations, grid switching stations and sub-stations all over the country wherever the supply voltage changes from one level to another.

The basic construction consists of three pairs of windings; one High Voltage, the other Low Voltage, mounted on separate limbs of a three-limb core. The coils are wound and distributed in the same manner as single phase windings but the level of insulation between windings will take into account the high voltages likely to be present.

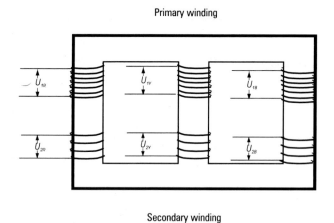

Figure 7.18   Core type three-phase transformer

The shell type has five limbs and is sometimes used for very large transformers.

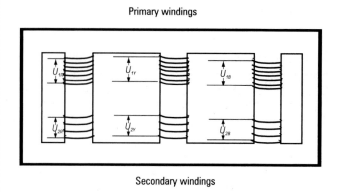

Figure 7.19   Shell type three-phase transformer

The transformer ratio rules will still apply but with a three phase transformer there is always the choice of STAR or DELTA connection.

The effect of this can be seen as follows.

Example
The transformer has a ratio of 13:1 and is supplied at 22.52 kV

The primary windings are DELTA connected and the LV side is connected in STAR.

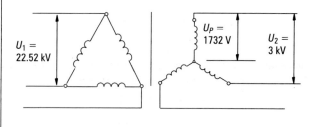

Figure 7.20

The voltage across each of the HV windings is 22.52 kV and assuming negligible losses, each LV winding will have a voltage of

$$\frac{22.52 \times 10^3}{13}$$

i.e. 1732.3 V

Since these windings are now to be connected in star connection the terminal voltage will be

$$U_2 \times \sqrt{3} = 1732.3 \times \sqrt{3} = 3 \text{ kV}$$

Now go to the Try this below.

Terminal markings

Transformer terminals are generally marked with CAPITAL LETTERS e.g. A1 B1 C1 on the HV side and lower case letters d2 e2 f2 on the LV side.

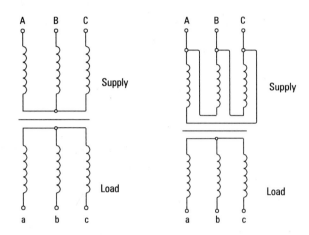

Figure 7.21

The windings themselves may connected in star or delta as required and the transformer may carry an additional marking to indicate that this has been done.

A transformer which is marked D.y is delta connected on the HV side (capital D) and star connected on the LV (lower case y).

Other variations would be   y.D ; D.d ;   Y.d or Y.y.

In addition to this, a number may be included in the code to indicate the phase relationship between the windings. This number is based on the clock face and uses the numbers on the dial rather than the phase angle.

For example, a transformer marked Y.y 6 is star connected on both HV and LV windings but the LV is 180 ° out of phase with the HV.

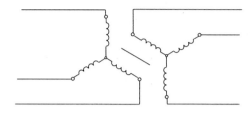

Figure 7.23   *Y-y 6 connected transformer*

Another may be marked Y.d 8 to indicate that the delta connected LV winding is 120 ° out of phase with the star connected HV side.

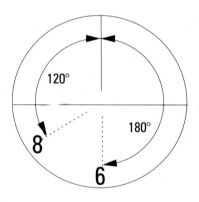

Figure 7.24

Try this
T_1 has a ratio of 8:1 and is connected star/delta to a 33 kV three phase supply.
T_2 has a ratio of 10:1 and is connected delta/star to the output circuit from T_1.
What is the voltage at the LV terminals of T_2?

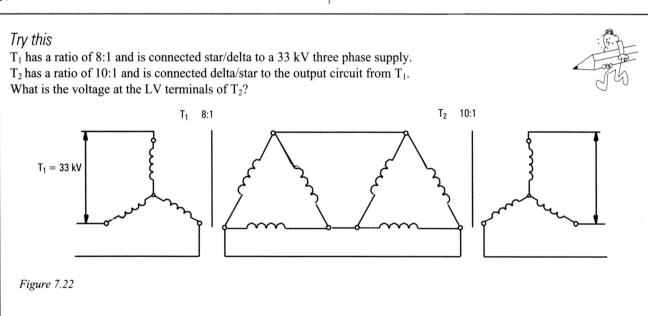

Figure 7.22

Where a transformer is working independently of others, the phase angle of the output voltage creates no particular problem as there is no inter-connection between this and any other circuits.

Where two or more transformers are connected to the same circuit, as in the case of ring main feeders, then the phase angle of the output voltage must be the same in all transformers.

Transformer efficiency

In practical terms it is clear that a transformer will not deliver exactly the same amount of power as it receives.

The efficiency of a transformer can be determined in a similar manner to efficiency calculations performed on any other form of energy-converting device.

$$\frac{\text{Output}}{\text{Input}} = \text{efficiency}$$

This is of course "per unit" efficiency and the result will be a fraction.

It is common practice to express efficiency as a percentage. This is nothing more than the per unit efficiency with the decimal point shifted two places to the right.

Example

A double wound transformer supplied with 16 A at 180 V supplies a load of 36 A at a terminal voltage of 75.2 V. What is the efficiency of the transformer?

$$\text{Efficiency} = \frac{\text{Power out}}{\text{Power in}}$$

$$= \frac{36 \times 75.2}{16 \times 180}$$

$$= \frac{2707.2}{2880}$$

$$= 0.94 \text{ p.u.}$$

or alternatively;

$$\% \text{ Efficiency} = \frac{\text{Power out} \times 100}{\text{Power in}}$$

$$= \frac{270720}{2880}$$

$$= 94\%$$

When transformer losses are taken into consideration the formula given below is used to calculate transformer efficiency.

$$\text{Efficiency} = \frac{\text{Output}}{\text{Output + losses}} \times 100\%$$

Transformer losses

A transformer is not 100% efficient because it has losses.

If one tenth of the power delivered to the transformer is taken up by the losses then only nine tenths of the input power would end up as output.

In other words

$$\text{Input} = \text{Output + Losses}$$

Transformer losses are generally grouped into two different categories.
- Copper losses and
- Iron losses

Example

The full-load copper and iron losses of a transformer are 15 kW and 10 kW respectively. If the full-load output of the transformer is 900 kW calculate the losses and the efficiency of the transformer on full-load.

Total loss $\quad$ = copper loss + iron loss

$\qquad\qquad$ = 15 + 10 = 25 kW

Efficiency $\quad = \dfrac{\text{Output}}{\text{Output + losses}} \times 100\%$

$\qquad\qquad = \dfrac{900}{900 + 25} \times 100$

$\qquad\qquad = 97.3\%$

Try this

The iron loss for a transformer is 6 kW and its full-load copper loss is 9 kW. If the full-load output is 500 kW calculate the total losses and the efficiency at full-load.

Copper losses

These are quite simply explained as the power lost through heat in a loaded winding.

Power is lost in a winding as a result of current passing through a resistance. The resistance of the winding will be determined by the length and cross-sectional area of the conductor. Nothing much can be done about the length because a transformer winding will need to have a pre-determined number of turns in order to meet its design specification. If however the cross sectional area is too small, then the efficiency of the transformer will suffer due to excessively high copper loss.

In terms of current and resistance we can express power loss as follows;

$\qquad$ Power $\quad = I^2 R$ Watts

From this it can be seen that power loss is proportional to the SQUARE of the current in the windings.

This is an important factor in transformer operation as any reduction in load current has a significant effect on copper loss, equivalent to the square of the fraction of full load current.

In other words:

A transformer at half full-load will have copper losses equivalent to one quarter of the full-load copper losses.

At one quarter of full-load, the copper losses will be one sixteenth of what they would have been if the transformer had been fully loaded.

By comparison, a transformer running at 10% overload would show a copper loss of 21% more than the design full-load value and consequently its efficiency would be adversely affected.

Example

A transformer has a full-load copper loss of 200 Watts. What is the copper loss, at
$\qquad$ a. two-thirds full load?
$\qquad$ b. one third full load?
a.

Copper loss $\quad$ = f.l. copper loss × the square of the
$\qquad\qquad\qquad$ fraction of full load

$\qquad\qquad = 200 \times (0.666)^2$

$\qquad\qquad = 88.7$ W

b.

Copper loss $\quad = 200 \times (0.333)^2$

$\qquad\qquad = 22.18$ W

Iron losses

The core material of a transformer must be chosen with great care in order to minimize the effects of iron loss. From your study of electromagnetic materials you will have learned that the core material will be subject to a certain amount of hysteresis loss (for revision see Chapter 1). This is the amount of energy input required to magnetise the core during the positive half cycle of the supply voltage then to de-magnetise and re-magnetise to the opposite polarity during the negative half cycle.

The hysteresis loss is proportional to the area enclosed by the loop and consequently an efficient transformer core material is one which is easily magnetised and de-magnetised and forms a narrow loop thus enclosing as small an area as possible. It must be remembered that the hysteresis loss is not a "one off" feature but is repeated with every cycle of the supply current.

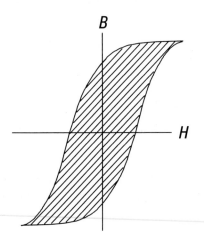

Figure 7.25   *High loss*

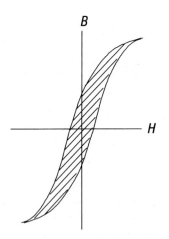

Figure 7.26   *Low loss*

Eddy current loss

Because of the nature of induced currents it is inevitable that alternating current will be induced in any conductor which is situated in an alternating magnetic field.

Since the core is likely to be made of ferromagnetic (iron based) material this is in itself a conductor material. Locally circulating (eddy) currents will be induced in the core material and unless prevented from doing so, these will circulate in the core performing no useful function and dissipate their energy in the form of unwanted heat thus reducing the overall efficiency of the device.

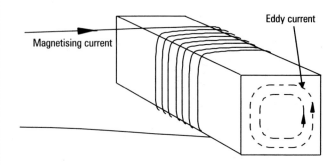

Figure 7.27

In a solid core, eddy currents will be induced and will flow freely in the core material.

If the core is made up of thin sheets or laminations as they are called, and each lamination is insulated from the next by a microscopically thin layer of insulating material then the eddy currents are virtually eliminated. The overall size of the core is only marginally greater than it would have been as a solid core and the flux path is not impeded.

Note: Iron losses do not change with a change in load.

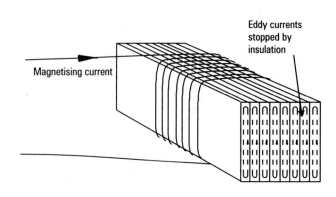

Figure 7.28

Other transformer types/terminals/efficiency and losses

The secondary of an isolating transformer has no physical electrical connection to the _____ supply and no connection to _____

If a transformer is assumed to be 100% efficient:

Power in = _____ .

An auto-transformer has only _____ _____ .

The core-type three-phase transformer has three _____ of windings; one _____ voltage, the other_____ voltage mounted on separate_____ of a three-limb _____ .

The shell-type three-phase transformer has _____ limbs.

Transformer terminal markings A1, B1 and C1 are on the _____ side and a1, b1, and c1 are on the _____ side.

When losses are taken into consideration the formula for transformer efficiency is:

Efficiency   =

Transformer losses are grouped into two different categories:

_____ losses which _____ when the load changes.

_____ losses which _____ change with a change in load.

Hysteresis loss is proportional to the_____ enclosed by the _____

A fat loop means _____ hysteresis loss and a thin loop means there is_____ hysteresis loss.

Part 3

Instrument transformers

Current and voltage transformers are a common feature of fixed panel meter installations. The use of instrument transformers brings several advantages to the user, some of which are;

- Increased safety
- Practicality
- Economy

Increased safety

Instrument transformers isolate the instrument circuit from the main circuit. This is an absolute necessity for high voltage installations but even in low voltage installations this isolation can improve safety. By earthing the secondary of the instrument transformer this ensures that the instruments are as close as possible to earth potential.

Practicality

Where very high values of current or voltage are to be indicated, the use of instrument transformers allows meters to be installed at any convenient position and not necessarily at the point of application. Meter leads may be as long as is necessary and of small cross section with light insulation.

Economy

The secondary values are a matter of choice and for economy's sake these are normally low values of current and voltage. The instruments installed are of conventional construction, readily available and do not have any expensive special features.

Current transformers

Current transformers, or CT's as they are commonly called, consist of a ring shaped core around which is wound several turns of wire which make up the secondary winding. The primary winding is usually a single conductor such as a large cross-section cable or bus-bar.

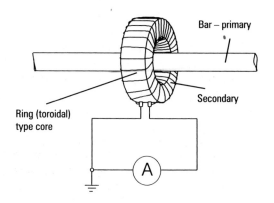

Figure 7.29   Bar primary current transformer

The secondary winding consists of a number of turns of insulated wire determined by the required output current.

For example if a current of 650A is to be indicated on a scale of 0 to 1000A using an ammeter whose full scale deflection is 5A then the CT would need to have an effective ratio of 1000 : 5. (or 200 : 1)

The current transformer ratio may, in some cases, be modified by passing the primary through the transformer more than once, as shown in Figure 7.30.

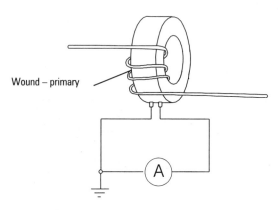

Wound – primary

Figure 7.30   Wound primary current transformer

This is clearly not practical where very large primary conductors are being used but the practice may be adopted with smaller cables.

The loading of a CT is known as the "burden". For example, a CT secondary may deliver 5 A at a voltage of 3 V thus giving a burden of 15 VA at an impedance of 0.6 Ω including the resistance of the connecting leads.

The impedance of the secondary circuit must be taken into account and can be calculated from

$$Z = \frac{U}{I}$$

The manufacturer will be able to give an estimate of the correct burden for a CT, but these are available in a range of values between 5 and 50 VA per phase.

CAUTION

A CT must never be left in place on a loaded conductor without a secondary load of some description. A loaded primary conductor always produces a secondary voltage, and without a secondary load this can be high enough to cause an accident or even damage the CT.

If the ammeter has to be removed or disconnected for any reason always short-circuit the CT terminals first before disconnecting it. This will not harm the CT in any way and will prevent a dangerous situation from arising.

Hint

Wrap several turns of bare copper wire around the CT terminals before disconnecting the meter. These can be left on and cut away with your side cutters after re-connection.

If a switching arrangement is used to distribute the CT secondary to several ammeters situated in different locations, the switch must be of the "make before break" variety so that the CT secondary is not broken during the switching operation.

Voltage transformers

Voltage transformers, often abbreviated to VTs or less commonly PTs (potential transformers) are used to connect a voltmeter or voltage coil to a mains supply. The use of the VT is essential for high voltage measurement such as 11 kV or 33 kV, but even at 400 V or 230 V the isolation from a direct mains connection and a possible reduction to 110 V can lead to increased safety.

A voltage transformer is a straightforward double wound transformer of conventional construction and is not to be shorted out when not in use. The secondary terminals should however be insulated and enclosed to prevent accidental contact.

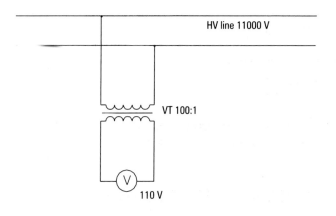

HV line 11000 V

VT 100:1

110 V

Figure 7.31

As in the case of the CT, a VT should be matched to its secondary circuit in terms of the appropriate "burden".

Burden

To calculate the burden of a measurement circuit, first note the current in the circuit, square it and multiply this by the impedance of the circuit. Or alternatively; measure the voltage across the transformer terminals and multiply this by the current in the circuit. In either case you will get a value in VoltAmps which should not be greater than the burden of the transformer.

Example

A CT has a secondary current of 15 A at an operating voltage of 2 V.

$$\text{Burden} = 15 \times 2 = 30 \text{ VA}$$

Try this

What is the burden of a CT which measures up to 5 A in a secondary circuit with an impedance of 1.2 Ω?

Installation and connection of switchgear

There are three main types of switchgear used in electrical systems and they are:

1. a circuit breaker
2. a switch
3. an isolator

The choice of switchgear for a particular application depends upon its function in the electrical system.

Function of switchgear

1. Circuit breaker (C.B.)

A circuit breaker is a mechanical device for "making" and "breaking" a circuit under **all** conditions (i.e. fault current and normal load current).

When a C.B. opens to interrupt a circuit an "arc" is drawn between the fixed and moving contacts. This is no real problem at domestic voltages, since the arc is small and is extinguished quickly in the atmosphere, but at extra high voltages the arc is much more difficult to extinguish and air or oil, often under pressure, have to be used.

air circuit breaker   – contacts "break" in air
oil circuit breaker   – contacts "break" whilst immersed in oil

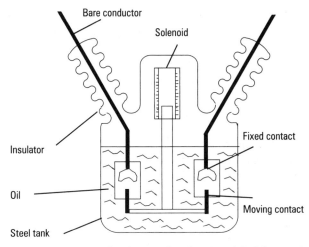

Figure 7.32   Bulk oil circuit breaker (simplified diagram)

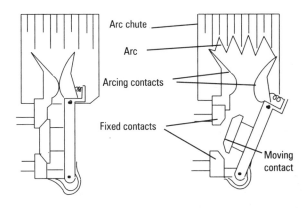

Figure 7.33   Air circuit breaker (simplified diagram)

Vacuum circuit breaker

A vacuum circuit breaker is simply a sealed switch with its fixed and moving contact enclosed in a high vacuum chamber, and is used on high voltage systems as an alternative to using oil or air circuit breakers.

Some vacuum circuit breakers are filled with sulphur hexafluoride (SF_6) gas under pressure. Its insulating and arc quenching properties are superior to air and oil. A vacuum C.B. is shown in Figure 7.34.

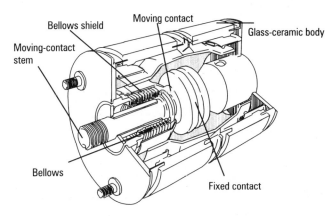

Figure 7.34   Vacuum circuit breaker

Air-blast circuit breaker

This type of circuit breaker uses compressed air to "close" and "open" the contacts. When the contacts are opening a jet of compressed air is directed across the path of the arc as it is formed and quickly extinguishes it.

2. Switch

A switch is a device for "making" and "breaking" a circuit under normal and overload conditions. It can "make", but not necessarily "break", a circuit under short-circuit conditions.

3. Isolator

An isolator differs from a switch in that it is intended to be "opened" when the circuit is not carrying current, however, an isolating switch may be used to close a circuit on to a load.

Typical applications of switchgear

Circuit breaker

- To connect a power station generator transformer to the 400 kV transmission system.
- To connect an industrial consumer's supply to the 33 kV or 11 kV distribution system (Figure 7.36).

Switch

- Domestic premises – for making and breaking the supply to an appliance, luminaire, heater and so on.
- Industrial premises – for starting and stopping an electric motor. D-O-L starters, which have an electrically operated switching device (called the contactor) are often used, or a manual starter could be used.

Isolator

- To isolate the consumer's electrical installation from the electrical supply.
- To isolate a motor circuit from the electrical supply.
- To isolate a high voltage circuit breaker from the electrical supply to enable work to be carried out on it (Figure 7.35).

Purpose of switching devices in compliance with BS 7671

Isolator

– to isolate the consumer's installation from **ALL LIVE** supply conductors

Emergency switch

– to cut off the supply rapidly from an installation to prevent or remove danger

Functional switch

– to control part of a circuit independently from other parts of the installation

Mechanical maintenance

– a means of switching off for mechanical maintenance shall be provided where mechanical maintenance may involve a risk of burns or a risk of injury from mechanical movement

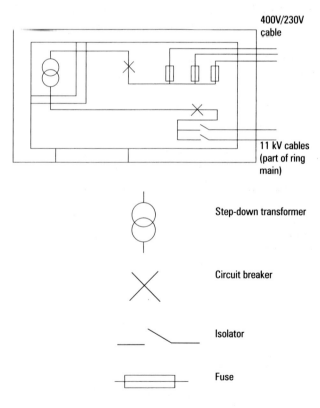

Figure 7.36   Small sub-station and associated switchgear (Schematic diagram)

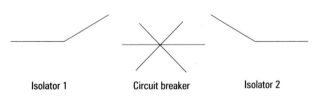

Figure 7.35

Instrument transformers/switchgear

Current transformers are commonly called _____ and they consist of a _____ shaped core around which is wound the _____ winding.

The primary may be a _____ primary or a _____ primary.

Before disconnecting an ammeter from the secondary of a C.T. _____ the terminals first.

The three main types of switchgear are:

1. _____

2. _____

3. _____

A C.B. is a mechanical device for _____ and _____ a circuit under _____ conditions.

A vacuum C.B. is simply a _____ with its fixed and moving _____ enclosed in a high _____.

Air blast C.B.s use _____ air to _____ and _____ the contacts.

A switch will _____ and _____ a circuit under _____ and _____ conditions.

An isolator should isolate the _____ installation from _____ supply conductors.

An emergency switch is used to _____ the supply _____ from an installation to _____ or _____ danger.

A _____ is used to control part of a circuit independently from other _____ of the installation.

1. Sketch the core arrangements for EACH of the following types of single-phase transformers:
   (a) core
   (b) shell
   (c) toroidal

2. Calculate the secondary voltage of a single-phase transformer with a turns ratio of 1:20. The primary winding has 115 turns and 2 V per turn.

3. Sketch the winding arrangements for EACH of the following types of single-phase transformer.
   (a) Double-wound step down transformer
   (b) Step-down auto-transformer
   (c) Step-up auto-transformer

4. A single-phase, 230 V/50 V transformer has a secondary current of 6 A. Calculate the current in the primary circuit.

5. Describe, with the aid of a diagram, the basic principle of operation of a single-phase step-up transformer.

6. State:
   (a) the reason for laminating transformer cores
   (b) the precaution that must be taken when disconnecting an ammeter from a current transformer

7. (a) State the purpose of a circuit breaker.
   (b) Give one application of a circuit breaker.
   (c) Name THREE types of circuit breaker.

8. State the purpose of THREE types of switching device to comply with Part 4 of BS 7671.

8

Illumination

Answer the following questions to remind yourself of what was covered in Chapter 7.

1. What is the turns ratio of a transformer which provides a 50 V output from a 230 V supply?

2. A single-phase transformer has an iron loss of 4 kW and a full-load copper loss of 10 kW. If the full-load output is 400 kW, calculate the total losses and the efficiency at full-load.

3. Transformer losses are generally grouped into two different categories:
   copper losses and iron losses
   (a) State which losses change with a change in load, and which losses remain the same.
   (b) What is hysteresis loss proportional to?
   (c) How are eddy currents produced?

4. Sketch simple diagrams of the following three-phase transformer core arrangements:
   (a) core type
   (b) shell type

On completion of this chapter you should be able to:

◆ identify the visual range of the spectrum
◆ state units, quantities and factors used in illumination?
◆ compare the efficacies and colour rendering properties of different light sources
◆ with the aid of diagrams perform simple calculations, related to illumination
◆ describe the construction, operation and application of incandescent and discharge lamps
◆ identify the control requirements of discharge lighting circuits

Part 1

Electric lighting has now been available for over 100 years. In that time it has changed in many ways and yet a lot of the same ideas are still in use. The first type of electric light was the arc lamp which used electrodes to draw an arc through the air. There were many accidents with these and regulations had to be brought in to control their use. The first lamp developed that was truly suitable for indoor use was the carbon filament lamp. Although by modern standards it was very dim it was cleaner and far less dangerous than the exposed arc lamp.

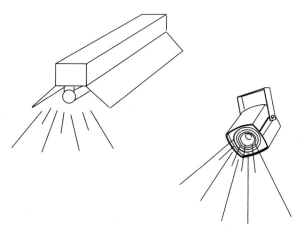

Figure 8.1

Visible light

When modern electric lighting installations are designed there are a number of factors that should be considered. One of these is the intensity of the light or how bright it seems to us. The human eye responds to a wavelength range of frequency which is between about 400 and 800 nanometres (nm).

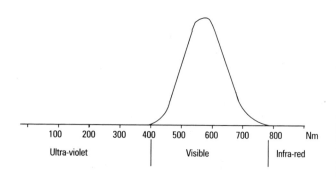

Figure 8.2

The visible range for most people goes between 380 and 780 nm.

Table 8.1

Colour	nm
VIOLET	380-436
BLUE	436-495
GREEN	495-566
YELLOW	566-589
ORANGE	589-627
RED	627-780

The various types of light sources use different parts of the spectrum and this can have an effect on their application. In areas where red food is on display it is not a good idea to have a light source that brings out green colours. A bright yellow lamp is not good if blues and reds are critical colours. So the colour rendering of the light source can be very important.

Lighting units and quantities

Although it is not an aim of this book to get involved with the theory of lighting design, there are some lighting terms that you will need to know and the units in which they are measured that you will need to recognise.

LUMINOUS INTENSITY is the light power of the source of illumination measured in **CANDELA**.

LUMINOUS FLUX is the light emitted by a source and is measured in **LUMENS**.

ILLUMINANCE is a measure of the density of luminous flux at a surface and is measured in **LUX** (lumens per square metre).

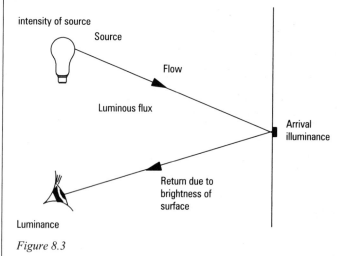

Figure 8.3

LUMINANCE is a measure of the light given off from a surface measured in **CANDELA PER SQUARE METRE**.

LUMINOUS EFFICACY is the ratio of the luminous flux emitted by a lamp to the power the lamp and control gear consume. This is measured in **LUMENS PER WATT** and can be used to compare different types of lamp.

Table 1.2

Quantity	Quantity symbol	Unit	Unit symbol
Luminous intensity	I	Candela	cd
Luminous flux	Φ	Lumen	lm
Illuminance	E	Lux	lx
Luminance	L	Candela per square metre	cd/m^2
Luminous efficacy		Lumens per watt	lm/W

Luminous efficacy

When this is being applied to discharge lighting it is not sufficient to take the lamp watts at their stated value. To calculate the efficacy it is necessary to take the lamp power and the control gear losses into account.

Luminous efficacy will be used throughout this book to compare the different types of lamp.

Examples of the different lamp efficacies are shown in Table 8.3.

Note: Efficacy replaces efficiency because it is very difficult to measure the light output in watts, but very simple in lumens.

Table 8.3

Type of lamp	Efficacy (lm/Watt)
(GLS) Filament	10–18
(TH) Tungsten halogen	12–22
(MPS) HP Mercury	32–58
(MCF) Fluorescent	60–78
(SON) HP Sodium	55–120
(SOX) LP Sodium	70–160

Calculating luminous efficacy

$$\text{Efficacy (1 m/W)} = \frac{\text{light output (lm)}}{\text{electrical input (W)}}$$

Example
Determine the efficacy of a 150 W tungsten filament lamp which gives an average light output of 1960 lumens.

$$\text{Efficacy} = \frac{1960}{150} = 13 \text{ lm/W}$$

Try this
An 80 W tubular fluorescent lamp has an average light output of 5000 lumens, determine the efficacy of the lamp.

Remember
The input to a lamp is measured in its electrical power unit, the watt, and the light output is measured in lumens.

Sources of lighting from electricity

There are several ways of producing light from an electrical source. These can be divided into two main headings:

1.   **Incandescent (glowing with heat)**
A fine filament of wire is connected across an electrical supply and is made to heat up until it is white hot and gives off light.

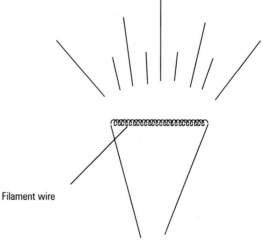

Filament wire

Figure 8.4   Incandescent

2. Discharge (arc)

Electricity is made to flow through a gas or vapour so that the atoms are agitated or excited. This atomic excitation produces light.

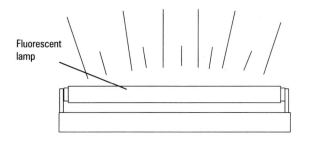

Figure 8.5

To improve the light output and colour rendering of lamps, phosphors are often used as a coating inside the lamp envelope. These phosphors react with the ultra violet energy emitted from the electrical discharge and convert it into visible light.

Reflectors and lenses

To further improve the light output reflectors and lenses can be used.

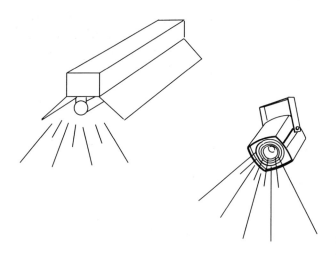

Figure 8.6

Luminaires

The light sources, although contained within their own envelope, are usually fitted in a luminaire. The luminaire can take almost any form and is usually designed for an application.

The function of most luminaires is to:
* redirect the light from the lamp with the minimum of loss
* reduce glare from the light source
* be acceptable for the environment they are to be installed in

Aspects that must be considered when selecting a luminaire can be listed as:
* the degree of lamp protection
* electrical safety
* heat dissipation
* the finish

There are a number of **SAFETY** labels that may assist when selecting lamps and luminaires. As all lamps give off some heat this must be considered so that the heat given off by the luminaire will not raise the temperature of the surface on which it is mounted above its rated limit.

Luminaires marked with the "F-Mark" symbol are suitable for mounting on any normal surfaces.

Figure 8.7   F-Mark

The reflected heat may also be a problem and a minimum distance may be recommended between the luminaire and the surfaces to be lit. This can be indicated by the symbol shown in Figure 8.8.

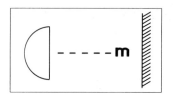

Figure 8.8   "m" denotes the minimum distance recommended between the luminaire and the lit surface.

When luminaires have to be installed in hazardous areas where there is a concentration of explosive gas or vapour then the appropriate enclosure should be used. To help to identify these the symbols in Figures 8.9 and 8.10 are used. The degree of protection must also be shown as it is very important to have the correct enclosure in the relevant environment.

Figure 8.9   BASEEFA Mark
Luminaires certified by the British Approvals Service for Electrical Equipment in Flammable Atmospheres carry the BASEEFA mark.

Figure 8.10   EEC Mark
For luminaires in hazardous areas

BSI Safety Mark

The safety mark is a trade mark affixed to products under licence from BSI. It demonstrates that the product meets the safety requirements of the relevant standards and that the manufacturer's quality system is assessed against ISO 9000 or a similar quality management standard.

Figure 8.11   BSI Safety Mark
Reproduced with kind permission from BSI

Points to remember ◀ – – – – – – – – – –

Lighting units and quantities/luminaires

The candela is the unit of _____ _____.

Luminous flux is measured in _____, and illuminance is measured in _____.

Lumens per watt is the unit of _____ _____.

The filament lamp has a _____ luminous efficacy, and the L.P. sodium lamp has a _____ luminous efficacy.

The phosphor coating which is applied to the inside of a _____ lamp envelope converts _____ light into _____ light.

Reflectors and lenses are used to improve the _____ _____.

Two functions of a luminaire are:
1.   they must redirect the_____ from the_____ with the minimum of _____.
2.   reduce the _____ from the light source.

Luminaires marked

are suitable for mounting on any _____ _____.

On the symbol

"m" denotes the _____ _____ recommended between the_____ and the _____ _____.

Part 2

Inverse square law

The illuminance on a surface which is produced by a single light source, varies inversely as the square of the distance from the source. This is known as the INVERSE SQUARE LAW and is shown by the expression:

$$E = \frac{I}{d^2}$$

where:

$E =$ illuminance in LUX

$I =$ luminous intensity of the light source in CANDELA

$d =$ distance in metres from the light source to a point on a surface

The INVERSE SQUARE LAW is illustrated in Figures 8.12 to 14. An incandescent lamp of luminous intensity 500 candelas is fixed at different distances above a flat surface. The value of illuminance E on the surface is calculated for each distance d.

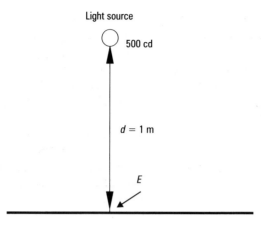

Figure 8.12

$$E = \frac{I}{d^2} = \frac{500}{1^2} = 500 \text{ lux}$$

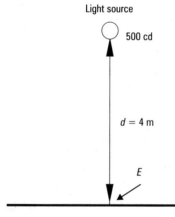

Figure 8.13

$$E = \frac{I}{d^2} = \frac{500}{2^2} = 125 \text{ lux}$$

Figure 8.14

$$E = \frac{I}{d^2} = \frac{500}{4^2} = 31.25 \text{ lux}$$

If the distance is doubled between the light source and surface, the illuminance E will fall to one quarter of the previous value. Although the illuminated area will increase in size, the illuminance on the surface however will decrease accordingly. This effect can be demonstrated by shining a torchlight directly onto a flat surface and observing the increased area of light when the torch is moved further away from the surface.

Try this
A 1500 cd lamp is suspended 3 m above a bench. Calculate the illumination at a point directly below the lamp.

Cosine law

If a beam of light from a lamp hits a surface at an angle, the illuminated area increases but the illuminance on the surface is lower than when the light is pointed directly at the surface. This effect can be demonstrated by holding a torchlight at an angle to a surface, rather than directly at the surface, and observing the increased illuminated area. The illuminance at a point on the surface will now be reduced by a factor of the cosine of the angle. This is known as the COSINE LAW and is shown by the expression:

$$E = \frac{I}{d^2} \times \cos\theta$$

The COSINE LAW is illustrated in Figure 8.15. A 500 cd incandescent lamp is fixed at a height of 2 metres directly above a long bench, and the value of illuminance at point P is to be determined.

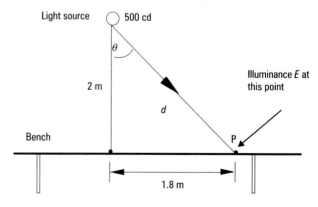

Figure 8.15

$$E = \frac{I}{d^2} \times \cos\theta$$

Distance d must be found:

$$d = \sqrt{2^2 + 1.8^2}$$

$$= \sqrt{4 + 3.24}$$

$$= \sqrt{7.24}$$

$$= 2.69 \text{ m}$$

Applying the Cosine Law:

$$E = \frac{500}{2.69^2} \times \frac{2}{2.69}$$

$$= 51.373 \text{ lux}$$
illuminance at P

Try this

Determine the illuminance at points P_1 and P_2 on the working plane.

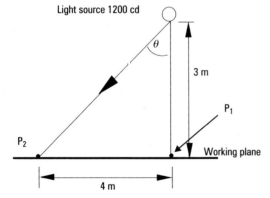

Light source 1200 cd

θ

3 m

P_1

Working plane

P_2

4 m

Figure 8.16

Maintenance factor (symbol M)

During the initial planning stages of a lighting scheme the designer must make allowances for the depreciation of illuminance due to dirt accumulating on the lamp, luminaire and room surfaces. A maintenance factor which is always less than unity is used in calculations to take account of the losses.

The degree of dirtiness will depend upon the type of premises and the nature of the work carried out within the premises. Maintenance factors in the range of 0.8 to 0.9 may be suitable where fairly clean areas are concerned. However, in foundries, welding shops and other dirty locations a maintenance factor of 0.65 would not be out of place.

Maintaining adequate levels of illuminance on the working surface will also depend upon the frequency of cleaning the luminaires, including lamps, and the surrounding walls and ceilings. Redecorating should also be carried out at regular intervals.

Light loss factor (LLF)

The maintenance factor (MF) must not be confused with the "light loss factor" which represents the total light depreciation at a given time compared to the figure when the installation was brand new in pristine conditions.

Coefficient of utilisation factor (symbol U)

The coefficient of utilisation factor is the proportion of the luminous flux which reaches the working plane. It is a measure of the efficiency with which light emitted from the lamp is used to illuminate the working plane.

Coefficient of utilisation factor

$$= \frac{\text{luminous flux reaching working plane}}{\text{total luminous flux emitted from luminare}}$$

Practical values for the C of U factor vary widely between about 0.1 (10% of emitted light reaches working surface) and about 0.95 (95%).

The C of U value depends on a number of factors. These are:
1. Type of luminaire (light fitting)
2. Colour and texture of walls and ceiling
3. Room size
4. Number and size of windows
5. Mounting height of luminaires

Considering the factors above in more detail:
1. Open type luminaires will allow more light to escape than enclosed types.
2. Light coloured smooth surfaces reflect more light than dark, matt surfaces.
3. A large room allows more light to be received directly from luminaires without reflection.
4. Uncurtained windows reflect virtually no light.
5. The higher the luminaire, the lower the illuminance on the surface below it.

Example

It is estimated that the lamps in an office reception area 7 m × 9 m emit 18000 lumens. Assuming a coefficient of utilisation of 0.7 and a maintenance factor of 0.9 calculate the illuminance.

Using the formula:

$$E = \frac{\Phi \times U \times M}{A}$$

Where

E = required illuminance (lx)
Φ = lumen output from lamps (lm), same as luminous flux
A = area to be illuminated (m^2)
U = coefficient of utilisation
M = maintenance factor

the answer is

$$E = \frac{18000 \times 0.7 \times 0.9}{7 \times 9}$$

$$= 180 \text{ lux}$$

Example

Calculate the luminous flux required to provide an illumination level of 150 lux in a room 5 m × 8 m if the coefficient of utilisation and maintenance factors are 0.6 and 0.8 respectively.

$$E = \frac{\Phi \times U \times M}{A}$$

$$\therefore \quad \Phi = \frac{E \times A}{U \times M}$$

$$= \frac{150 \times 5 \times 8}{0.6 \times 0.8}$$

$$= 12500 \text{ lumens}$$

Note: Light loss factor is sometimes used in these calculations instead of maintenance factor.

Remember

Not only does the decor of a room deteriorate with age, the lamps and luminaires themselves deteriorate with age (for example the light output from a fluorescent tube gradually becomes dimmer and luminaire reflectors and diffusers gradually become discoloured due to the heat produced by the lamp) therefore the maintenance factor and the light loss factor will be affected by these deteriorations with aging.

Try this

Calculate the total luminous flux required to provide an illumination level of 120 lux in a room 6 m × 8 m if the utilisation and light loss factors are 0.65 and 0.85 respectively.

Points to remember ◄---------------

Inverse square/cosine laws/lighting factors

When applying the _____ _____ Law the formula shown below is used,

$$E = \frac{I}{d^2}$$

where

E = _____
I = _____
d = _____

The _____ Law is represented by the expression,

$$E = \frac{I}{d^2} \times \cos\theta$$

L.L.F stands for _____ _____ _____ .

Maintenance factor is affect by the deterioration of the _____ itself, and the deterioration of the _____ of the room.

The coefficient of _____ factor depends upon
1. the type of _____
2. the _____ and texture of _____ and ceilings
3. the _____ size
4. the _____ and _____ of windows
5. the _____ height of the luminaires

Part 3

Incandescent lamps

Lamp designation
These are the abbreviations used to define types of incandescent lamps:

GLS   General lighting service

TH   Tungsten halogen

PAR   Followed by the lamp nominal diameter in eighths of an inch

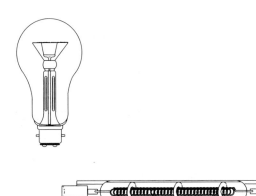

Figure 8.17

Incandescent lamps
Light is created in an incandescent lamp by heating a filament to a temperature of about 2800 K (2500 °C). The proportion of electrical energy converted into light is very small as most of it is converted into heat as infra-red energy. The light is mainly towards the red end of the visual spectrum which gives an overall warm appearance.

We shall now look at two incandescent lamps in more detail:
1. The GLS or general service lamp
2. The tungsten halogen lamp

The GLS lamp

This is commonly referred to as the light bulb.

The construction of this type of lamp is basically that shown in Figure 8.18 There are slight variations for different characteristics.

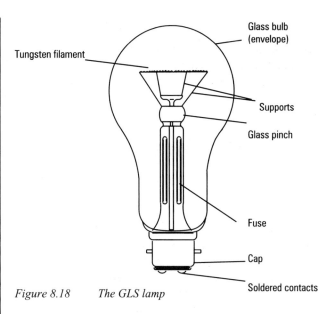

Figure 8.18   The GLS lamp

A current is passed through the tungsten filament which heats up. The light given off changes from a red colour at low temperatures to a whiter colour as the temperature increases. This can be seen by connecting a clear GLS lamp to a variable voltage supply. When the voltage is low the filament glows red and as the voltage is increased the colour changes until at full voltage the filament is a bright white.

As tungsten is used as the filament, and this melts at 3380 °C, the operating temperature of the GLS filament lamp is kept around 2500 °C. If a tungsten filament was taken up to this temperature in air it would tend to evaporate. To reduce the possibility of this happening the oxygen is pumped out of the glass envelope. On low power lamps such as 15 watt and 25 watt the area inside the glass bulb remains as a vacuum but on lamps of 40 watts and above the bulb is gas filled.

The gases most commonly used are argon and nitrogen. Argon to reduce the evaporation process and nitrogen to minimise the risk of arcing.

The filament is usually constructed of a number of coils of fine tungsten wire. If a single straight wire was used the gas would circulate freely around it and reduce its working temperature. By coiling the wire the heat radiated between the adjacent turns raises the temperature of the filament and improves the efficiency. By further coiling it, creating a coiled coil, the efficiency can be increased by a further 15%.

The efficacy of GLS lamps is between 10 and 18 lumens per watt. As this is low compared with other types of lamp its use can be limited. However it is the most familiar type of light source and has many advantages, including:
- comparatively low initial cost
- immediate light when switched on
- good colour rendering
- no control gear
- they can be easily dimmed

Their average life is 1000 hours.

The tungsten halogen lamp

As we have established tungsten filaments evaporate unless special conditions are created.

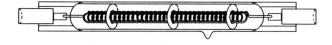

Figure 8.19   Linear tungsten halogen lamp

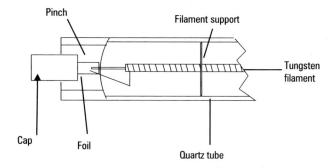

Figure 8.20

In the tungsten halogen lamp the glass bulb of the GLS is replaced with a small quartz envelope (as shown in Figures 8.19 and 8.20). Quartz will operate at higher temperatures than glass and the pressure inside the lamp can be considerably increased. This has the effect of slowing down the tungsten evaporation and can improve lamp life and efficiency. The most important design aspect of a tungsten halogen lamp is the introduction of a small quantity of halogen in the gas filling. The halogen elements used in these lamps include iodine, bromine and chlorine. These have the ability to combine with atoms of other elements without altering their general characteristics.

Under the right conditions a process known as the halogen cycle can take place. The conditions are critical, for example the temperature of the quartz envelope must be above 250 °C. As the tungsten gets very hot some evaporation takes place. The halogen will combine with the evaporated tungsten near the filament to form a metal halide. This moves across to the quartz envelope but will not be deposited on it due to its temperature. When the metal halide comes closer to the much hotter filament it decomposes; the tungsten is deposited on the filament and the halogen is released. This cleaning cycle repeats continuously returning the evaporated tungsten back to the filament, as shown in Figure 8.21.

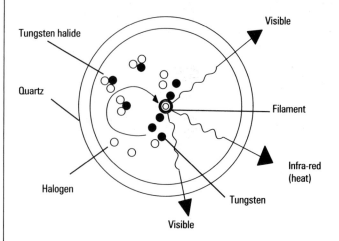

Figure 8.21   Enlarged cross-section

Remember
Tungsten lamps get very hot and need to be kept away from combustible material.

The efficacy of tungsten halogen lamps range from 12 to 22 lumens per watt which is generally higher than for the GLS lamp. These lamps give immediate light with good colour rendering and a warm colour appearance. They have many applications including:

- display lighting – as they can produce a "sparkle" effect
- exhibitions
- photographic
- security and floodlighting

Extra low voltage tungsten halogen lamps

These are produced for use not only on mains supply voltages but also on extra low voltages. This has meant that they are used extensively in the automobile industry for vehicle headlamps, as shown in the example in Figure 8.22.

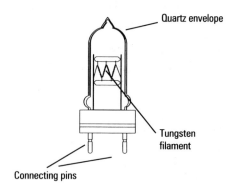

Figure 8.22

They can be combined with a precision faced glass reflector for display spot lights, as shown in Figure 8.23. The supply for these may be from an inbuilt 230 V to 12 V transformer or a separate 12 V supply.

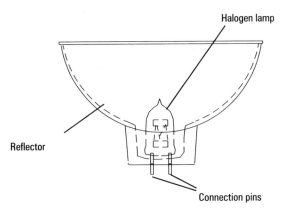

Figure 8.23

Lamp connections
The high operating temperature that they require to work can be a limiting factor in their use. The connection to the filament can be as shown in Figure 8.19 at each end of the lamp or as shown in Figures 8.22 and 8.23.

Lamp replacement
Contamination such as grease from finger marks can considerably shorten the life of a tungsten halogen lamp. New lamps are packed so that they can be installed without touching the quartz. Should they become contaminated they should be cleaned with a spirit before they are used.

Operating positions
Some low wattage lamps can be operated in any position. However, linear lamps should be fitted at or within 10° of horizontal, so that the heat is evenly distributed.

Remember
Incandescent lamps all work on the principle of a continuous tungsten filament that heats up and glows white.

Tungsten halogen lamps work at very high temperatures and require special consideration.

The high bright white light given from the tungsten halogen lamp makes it ideal for many display applications.

Discharge lighting
When an electric current is passed through a gas or metallic vapour it is said that an electrical discharge takes place. If this discharge is contained within a glass or quartz envelope then it can be used to produce light. The discharge, which is an electrical arc, "excites" or transfers energy to the gas or metallic vapour. This results in energy being emitted at certain characteristic wavelengths. It depends on the elements present in the discharge together with the temperature and pressure of the arc atmosphere, as to the precise energy wavelengths emitted. These wavelengths may be in the visual spectrum or in either the ultra-violet or infra-red regions.

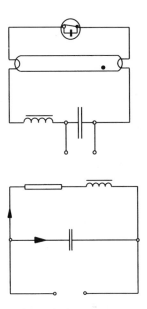

Figure 8.24   Discharge lighting circuits

Discharge lighting control gear
All discharge lighting circuits require some control gear to help them strike and then keep them working efficiently once they have struck.

Two of the main components used in these circuits are:
- the inductor (choke or ballast)
- the power factor correction capacitor

The inductor (choke or ballast)
An inductor is basically a coil of wire. In the choke or ballast for a discharge lighting circuit the wire will be wound round an iron core.

The choke or ballast has two basic functions:
- to initiate the discharge in the lamp i.e. to cause the electrical arc in the lamp to strike
- to limit the current through the lamp once the arc is struck

The choke and ballast both work on the principle of electromagnetic induction (remember this was covered in Chapter 1) and it is when the inductive circuit is actually broken that it causes a high voltage surge across the lamp which is sufficient to strike the main arc in the lamp.

153

The power factor correction capacitor

A discharge lamp will operate without a power factor correction capacitor, however, the power factor of the circuit would be very poor. To improve this poor power factor a p.f. correction capacitor is connected across the supply of the luminaire and its the leading power factor of the capacitor which counteracts the lagging power factor of the inductor (choke or ballast) – remember this was covered in Chapter 4.

Switches for discharge lighting circuits

As we have seen discharge lighting circuits are inductive and can cause excessive wear on the functional switch contacts (due to contact arcing, mainly when opening the inductive circuit).

When choosing a switch for a discharge lighting circuit, if it is not designed for switching inductive loads, it is recommended that it should be rated at twice the total steady current of the circuit, i.e. a 5 A circuit would require a 10 A functional switch.

Points to remember ◀ – – – – – – – – – – –

Incandescent lamps/discharge lighting

The glass bulb of a tungsten filament lamp, of 40W and above, is _____ filled. Argon to reduce the _____ of the filament and nitrogen to minimise the risk of _____ .

Tungsten halogen lamps work at _____ _____ temperatures and the halogen maintains the _____ _____ cycle and prolongs the _____ of the filament.

The _____ _____ _____ light from this type of lamp makes them ideal for _____ applications.

When an electric current is passed through a gas or _____ vapour an electrical _____ takes place.

The precise energy wavelengths emitted depend upon the elements present in the _____ together with the _____ and _____ of the arc atmosphere.

The choke, which is a type of _____ , has two basic functions:
1. to initiate the _____ in the lamp
2. to _____ the current through the _____ once the arc is _____ .

Part 4

Mercury vapour lamps

There are two types of mercury vapour lamps:
- Low pressure mercury vapour lamps
- High pressure mercury vapour lamps

Lamp designations
MCF
> Low Pressure Mercury – fluorescent lamp

MBF
> High Pressure Mercury with phosphor coating

MBI
> High Pressure Mercury with metallic halides

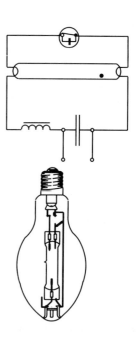

Figure 8.25

Low pressure mercury vapour lamps (fluorescents)

The low pressure mercury vapour lamp is more commonly known as the fluorescent lamp. In these the discharge arc takes place in a clear glass tube internally coated with a phosphor powder. The electrical discharge produces mostly ultra-violet and some blue and green light. The light output from the tube is produced by the phosphor coating when it converts the ultra-violet energy produced by the discharge into light.

Cathodes, coated with an electron emitting material are sealed into each end of the glass tube and connected to the pins of the lamp caps. Figure 8.26.

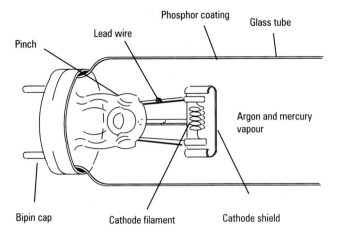

Figure 8.26

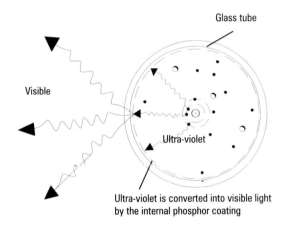

Figure 8.27

The lamp contains gases such as argon and krypton with a drop of liquid mercury. Small quantities are present so the gas pressure in the tube is low, about $\frac{1}{200}$th of atmospheric pressure.

In general a control circuit is required which first causes an electric current to flow through each cathode. The heated cathodes emit a cloud of electrons which are negatively charged. The electrons then have to be accelerated through the discharge tube by a high voltage being applied across it. This sustains the discharge and the excitation of the mercury atoms. The current flow through the lamp is then limited by the external control circuit.

As the control equipment also requires some current to operate it the power in a fluorescent circuit is always greater than the lamp rating. The manufacturer will usually give details of what the total rating is but if this information is not available the lamp current rating should be multiplied by a factor of 1.8.

There are a number of different control circuits, four of which are:

- switch start
- semi-resonant
- lead-lag
- high frequency

Each of these will now be dealt with in turn.

> *Remember*
> Disposing of unwanted fluorescent tubes must be carried out so as not to harm people or the environment.
>

Switch start circuit

The circuit for the switch start fluorescent is as shown in Figure 8. 28.

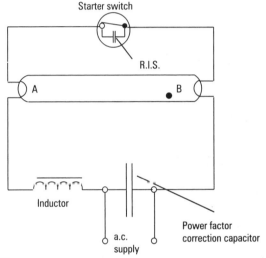

Figure 8.28

Note: the capacitor inside the starter switch is the radio interference suppressor.

When the circuit is first switched on current flows through the inductor, filament A, the gas in the starter switch, filament B and back to the supply. As the gas in the starter switch heats up, the bimetal contacts close. The resistance of the contacts is lower than that of the gas so the gas no longer conducts the current and it cools down causing the bimetal contacts to open. This momentarily switches off the current flow and a high voltage is induced across the lamp due to the sudden collapse of the magnetic field in the inductor.

The high voltage creates a discharge across the lamp and a new circuit is now completed shorting out the starter switch. The discharge across the lamp has a very low resistance and if the lamp was connected directly across the supply a very large current would result. The inductor is now in the circuit to see that this does not happen and limits the amount of current that is allowed to flow. The lamp will now continue to give off light until the circuit is broken or the supply is switched off.

The circuit described uses a "glow" type thermal starter which needs to be replaced from time to time. This type of circuit has the advantage that these starters are comparatively low in cost. There are however several disadvantages. These include:

- access necessary to replace the starter switch
- repeat starting cycles if the lamp does not start
- when the starter fails repeated starting attempts shorten the life of the other control equipment

Electronic start circuits

Many of the disadvantages of the glow starter switch circuit can be overcome by using electronic starter circuits, as in Figure 8.29.

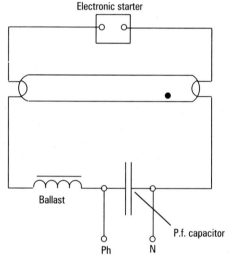

Figure 8.29

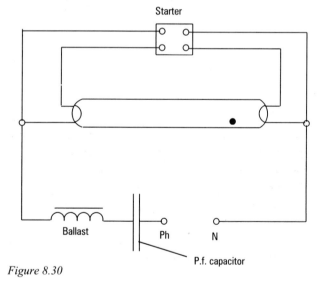

Figure 8.30

Electronic starters can provide a faster and flicker free start for the lamp. Many also contain a cut-out circuit which isolates the lamp in the event of a lamp failure. This is designed to extend the life of the ballast. The electronic starter has the advantage of having no moving parts and so the possibility of mechanical failure does not apply.

Semi-resonant circuits

There are many semi-resonant circuits in use by different manufacturers. The electronic start circuit shown in Figure 8.30 is an example of just one. As we have already seen a high voltage is required across the fluorescent tube to make the initial discharge. The semi-resonant circuit uses the high voltage created in an a.c. series circuit. The control circuit with the tube make up an R, L and C series circuit. Figure 8.31.

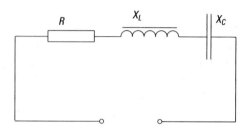

Figure 8.31

With true resonance X_L will equal X_C and so the only current limiting device will be the resistor. The danger is that the current will now rise to unmanageable limits. It does however mean that as the current increases the voltages across the inductor and capacitor also increase to very high values. Under these conditions it is possible to get thousands of volts across the inductor even though the supply voltage is 230 V a.c. By controlling the circuit it is possible to get a semi-resonance condition which keeps the current within manageable limits but gives high voltages which can be directed across the fluorescent tube. When the tube has struck it shorts out part of the ballast unit and the semi-resonant condition no longer exists.

By controlling the current in this way no extra power factor correction is required. It is important that should it become necessary to replace components, such as the capacitor (Figure 8.32), exact replacements are used. If components with other values were to be connected into the circuit the same semi-resonant conditions may not exist.

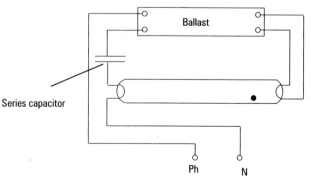

Figure 8.32

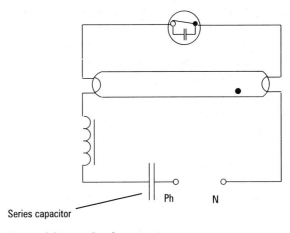

Series capacitor

Figure 8.34   Leading circuit

Lead-lag circuits

Where rotating machinery is used in an area lit by discharge lighting it is possible to get a stroboscopic effect. This means that the rotating parts may appear to be stationary even though they are still moving at high speeds. This effect is created by the flicker of the discharge lamp when it effectively switches off 100 times each second.

When a 3 phase supply is available adjacent single phase luminaires may be connected across different phases and neutral so that the lamps flicker at different times and reduce the overall effect. Where this is not possible twin luminaires with lead lag circuits can be used. A lead-lag circuit is, as the name implies, a circuit that contains one lamp that leads the other, hence the other lags. The lead-lag effect is produced by using the leading current effect of a capacitor and the lagging current effect of an inductor.

The lagging effect is produced naturally when an inductor is used in the circuit, as in Figure 8.33. The leading circuit uses a series capacitor which has a greater effect than the inductor in the circuit, as in Figure 8.34. When these two circuits are combined within a single luminaire there is no need for further power factor correction as one circuit will "correct" the effect of the other.

High frequency circuits

Whereas standard fluorescent circuits work on the mains supply frequency of 50 Hz, high frequency circuits operate at about 30000 Hz (30 kHz).

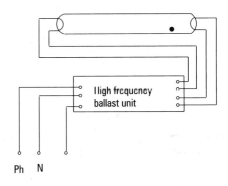

High frequency ballast unit

Ph   N

Figure 8.35

There are a number of advantages of high frequency circuits over standard ballast unit circuits. These include:

- higher lamp efficacy
- fast, first time, starting
- ballast shut down automatically for lamp failure
- no stroboscopic effect
- "noise" free
- can be regulated and controlled

A disadvantage is that it is important that supply cables within luminaires do not run adjacent to leads connected to the ballast output terminals as interference may occur. Also that the initial cost of these luminaires is greater than switch start circuits.

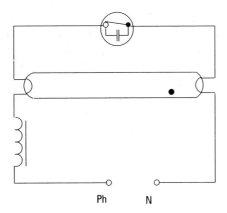

Ph   N

Figure 8.33   Lagging circuit

High pressure mercury vapour lamp (HPMV)

These lamps are used in industry and for outside lighting where colour rendering is not an important factor.

The high pressure mercury vapour discharge takes place inside a quartz arc tube. This is inside a bulb coated with a phosphor which converts ultra-violet radiation from the arc into light.

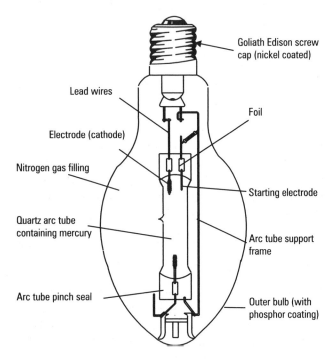

Figure 8.36

As the electrodes are not pre-heated, in this type of lamp, their construction is solid rather than coiled wire. An electron emissive material is built into the electrode and held there by a coil of wire.

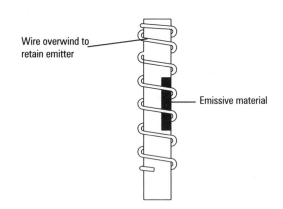

Figure 8.37   *Electrode (cathode) for HPMV lamp*

The starting of HPMV lamps is assisted by the use of an auxiliary electrode placed very close to a main electrode. (Figure 8.38)

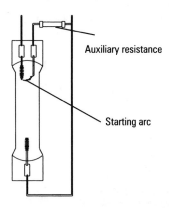

Figure 8.38

The auxiliary is connected via a resistance to the main electrode at the opposite end of the arc tube. When first switched on a small discharge is initiated between the main and auxiliary electrodes. This raises the temperature of the emissive material. The current from the discharge passes through the auxiliary resistance causing a voltage drop. The starting arc then collapses and this causes a high voltage to be induced from the choke. As one main electrode is now in an emissive state the main discharge is triggered off across the tube.

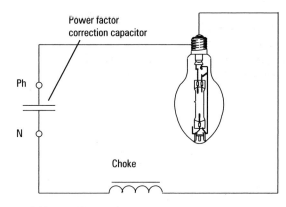

Figure 8.39   *Circuit diagram*

The lamp then builds up energy and in a few minutes gives off its full light. As with low pressure mercury circuits the choke becomes a current limiting device as soon as the arc has struck.

When the supply to the lamp is switched off it can take several minutes before it is back to full light. Before the arc can reform the temperature of the lamp must drop and decrease the internal pressure in the arc tube. When the pressure has dropped the starting cycle will begin again.

From the spectral distribution diagram it can be seen that the colour distribution is not regular. This can be improved in two ways:

- by coating the outer bulb with different phosphor powders
- by adding halides into the discharge tube

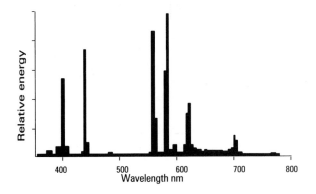

Figure 8.40   *Spectral distribution MBF*

Most HPMV lamps are coated with phosphor powders in much the same way as fluorescent tubes. It is possible to adjust the powders for different applications.

Those parts of the spectrum that are deficient in the arc of a HPMV lamp can be improved by incorporating other metals in the form of halides (halogen group includes – bromine, iodine, chlorine and fluorine).

There are many applications for HPMV lamps such as high bay lighting in industrial premises and sports centres, and street lighting. Halide lamps are also used for marshalling yards, shipping terminals and football stadiums.

Remember

The low pressure mercury vapour (fluorescent) lamp is by far the widest used type of discharge lamp.

There are several different circuits used to start and control the voltage and currents within the lamps and the application may determine their use. When using discharge lighting adjacent to rotating machinery the stroboscopic effect must be considered.

High pressure mercury vapour lamps are used mainly for outside lighting but can also be installed in industrial situations. Their disadvantage is that of taking time to get up to full brilliance both from start up and re-start after a short break in supply.

Points to remember ◄ – – – – – – – – – – – – – –

Mercury vapour lamps and circuits

The low pressure mercury vapour lamp is more commonly known as the _____ _____ .

A phosphor coating is applied to the inside of a fluorescent tube to convert the _____ light produced into a _____ light.

A switch start fluorescent circuit has a small capacitor inside the starter, which is for _____ _____ _____ , and a large capacitor across the supply, which is for _____ _____ _____ .

The discharge across a fluorescent tube has a very _____ resistance and if the tube was connected directly across the supply a _____ current would flow. The_____ prevents this situation from occurring.

Electronic starters can provide a _____ and _____ free start for a fluorescent lamp.

Three methods of overcoming stroboscopic effect are:
1. to connect fluorescent luminaires to _____ phases of the_____ _____ supply
2. to use _____ fluorescent luminaires with _____ circuits
3. to use _____ frequency fluorescent circuits

High pressure mercury vapour lamps will only restrike and return to full light output, after a supply failure, when the _____ of the lamp has fallen, which in turn decreases the internal _____ in the _____ tube.

Part 5

Sodium lamps

As with mercury lamps there are low and high pressure sodium lamps.

Lamp designations

SOX   Low pressure sodium lamp
                 - single ended

SON   High pressure sodium lamp

SON-T   High Pressure sodium lamp
                 - clear, single ended

SON-TD   High Pressure sodium lamp
                 - clear, double ended

Lamp operating positions

/V	Vertical – cap up
/D	Vertical – cap down
/H	Horizontal

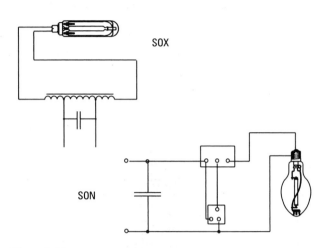

Figure 8.41

Low pressure sodium lamps

These lamps can be easily recognised by their distinctive bright yellow colour. As they have almost a single wavelength of 589nm which corresponds approximately to the point of maximum eye response they appear very efficient. Efficacies of over 160 lumens/watt are typical.

As this lamp has a monochromatic light output its colour rendering properties are virtually non-existent. This restricts its use to floodlighting or street lighting.

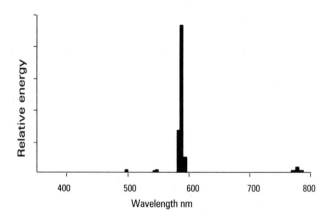

Figure 8.42   Spectral distribution SOX

The lamp consists of a long arc tube made of a type of glass construction known as ply tubing. As hot sodium vapour attacks any glass with more than a small amount of silica in it special glasses have been developed. These low silica glasses

are unfortunately expensive and difficult to work with. They have the added problem of being attacked by moisture. To overcome these problems ordinary soda-lime glass tubing has been produced which has an inner coating of the resistant glass. Hence it is known as ply tubing.

The tube is formed into a "U" shape and filled with a mixture of neon and 1% argon to between $\frac{1}{50} - \frac{1}{100}$th of an atmosphere. The lamp also contains a small quantity of metallic sodium which is solid at room temperature. Small protrusions are moulded in the arc tube for the sodium to collect in.

Each end of the discharge tube is a tungsten coiled coil electrode similar to those in fluorescent tubes. These have an emissive coating of barium oxide or similar material. As the electrodes are not used for heating their two ends are connected together.

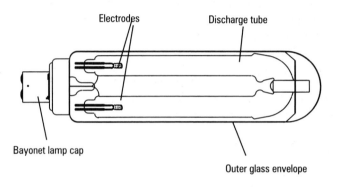

Figure 8.43   SOX lamp

An outer glass envelope is used to enclose the discharge tube. The space inside is fully evacuated. As the temperature of the arc tube must be maintained at about 270 °C to vaporise the correct amount of sodium the outer envelope has an inner reflective surface. This allows light to be transmitted but reflects the heat back onto the lamp.

Figure 8.43 shows the basic lamp construction.

It depends on the lamp rating as to whether a leaky transformer or ballast and igniter are used to make up the control gear, as shown in Figures 8.44 and 8.45.

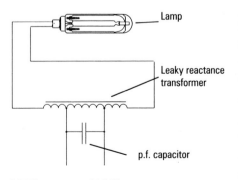

Figure 8.44   SOX lamps over 100 W

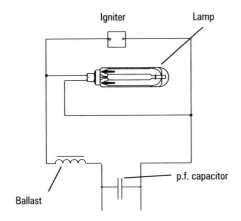

Figure 8.45   SOX lamps under 100W

High pressure sodium lamps

The high pressure sodium lamp combines the high efficacies of up to 112 lumens/watt with a good colour rendering. Their applications include high bay internal lighting and outdoor industrial or road lighting. It is also suitable for lighting sports centres and arenas but the yellow tint it gives means it is not suitable where colour television coverage is required.

Where external brick or stone surfaces are to be lit high pressure sodium is ideal as it brings out the stone's natural effect.

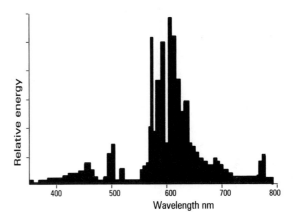

Figure 8.46   Spectral distribution SON

When a low pressure sodium lamp is first switched on a discharge occurs due to the mixture of neon and argon gases. This produces a distinctive red neon glow. The sodium now gradually becomes vaporised by the heat from the discharge and the sodium vapour slowly takes over. By using the neon, argon gas mixture the starting voltage can be reduced by over 30%. This makes the starting voltage almost independent of the ambient temperature and means that the lamp can be re-struck when hot within about one minute.

The lamp consists of a discharge tube of a sintered aluminium oxide (sintered alumina). This is a translucent ceramic material capable of operating at temperatures up to 1500 °C and withstanding hot sodium vapour.

Try this
Low pressure sodium lighting is distinctive by its bright yellow colour. List below applications that you can find where this type of lighting is in use.

Remember
A lamp (light source) which has the ability to show the true colours of objects has very good colour rendering properties.

The ends of the arc tube are hermetically sealed with sintered ceramic plugs containing electrodes. The electrodes are similar to those used in the HPMV lamps.

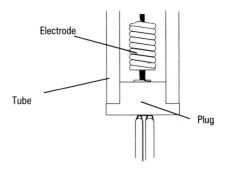

Figure 8.47   Discharge tube end

The arc tube usually contains sodium doped with mercury and argon or xenon. An outer bulb of weather resistant glass is used to protect the discharge tube. So that the arc tube temperature is maintained at not less than 750 °C the outer bulb is completely evacuated.

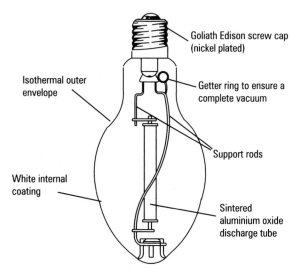

Goliath Edison screw cap (nickel plated)

Isothermal outer envelope

Getter ring to ensure a complete vacuum

Support rods

White internal coating

Sintered aluminium oxide discharge tube

Figure 8.48   SON

When switched on a pulse of between 2 kV and 4.5 kV is required to ionise the argon (or xenon). As this initial discharge heats up the sodium vaporises and then takes over the discharge. Without the mercury in the discharge the arc voltage would be as low as 50–60V but the mercury adds impedance and the voltage is around 150–200.

SON – starting methods

There are two types of lamp with regard to starting. Those that use an external starter in the control circuit, as in Figure 8.49 and those that contain a thermal starter inside the lamp, as in Figure 8.50.

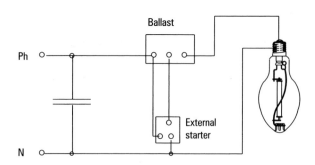

Ballast

Ph

External starter

N

Figure 8.49

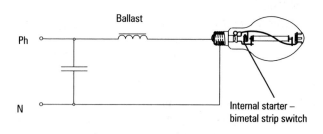

Ballast

Ph

N

Internal starter – bimetal strip switch

Figure 8.50

The high pressure sodium lamp does not have such a high efficacy as the low pressure type lamp. It does however have other advantages. For a start it offers a much improved colour rendering. The arc length is far shorter on the high pressure type and this makes it far more suitable for accurate optical control. In general a hot lamp, once extinguished, will not restart until cool, but some manufacturers are now using "hot start" igniters for immediate starting. The disadvantage of these is that they use pulses of about 9 kV.

Try this
The high pressure sodium lamp gives off a mellower yellow than the low pressure type. List below applications that you can find for the SON type lamp.

Points to remember ◀ – – – – – – – – – – – –

Sodium lamps and circuits

Low pressure sodium lamps are easily recognised by their distinctive _____ _____ colour.

The colour rendering properties of this type of lamp are very _____ compared to that of a high pressure sodium lamp which are very _____.

Low pressure sodium luminaires are used extensively for _____ _____.

When the low pressure sodium lamp is first switched on a _____ occurs due to the mixture of _____ and _____ gases, and then the sodium _____ by the heat from the discharge, and slowly takes over.

The efficacy of a low pressure lamp is _____ than the high pressure type.

Self assessment short answer questions
1. State the units which the following are measured in:
   (a) luminous intensity
   (b) luminous flux
   (c) efficacy
   (d) illuminance

2. Place in order of highest efficacy (lumens/watt) the following types of lamp.
   coiled coil filament
   low pressure sodium
   low pressure mercury vapour

3. An 800 cd lamp is suspended 2 m above a workbench. Calculate the illumination at a point directly below the lamp.

4. Calculate the illuminance at points P_1 and P_2 on the working plane.

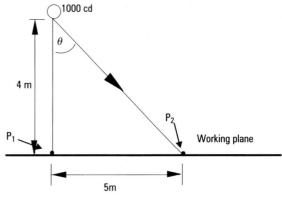

Figure 8.51

5. State three factors that will affect the coefficient of utilisation.

6. Explain briefly the function of EACH of the following components in a switch start fluorescent luminaire.
   (a) capacitor
   (b) choke
   (c) glow starter

7. State why:
   (a) tungsten halogen lamps contain halogen
   (b) fluorescent tubes are internally coated with phosphor powder
   (c) a high pressure mercury vapour lamp will not restart/strike immediately after an interruption of the supply.

8. Describe the starting sequence of a low pressure sodium lamp.

9

Efficiency, Work, Energy and Power

Answer the following questions to remind yourself of what was covered in Chapter 8.

1. State the lighting quantity which EACH of the following symbols represent:
   (a) I
   (b) E
   (c) Φ

2. Give THREE advantages of a GLS type incandescent lamp.

3. The circuit diagram below is for a low pressure mercury vapour lamp.
   (a) Identify the components marked A, B, C and D.
   (b) State ONE function of component A.

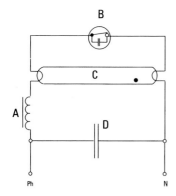

Figure 9.1

4. Calculate the luminous flux required to provide an illumination level of 180 lux in a room 5m × 7m if the maintenance and utilisation factors are 0.8 and 0.6 respectively.

On completion of this chapter you should be able to:

◆ define the terms efficiency, work, energy and power
◆ state the units for work, energy, power and the ratios applied for efficiency
◆ explain energy in terms of input and output energy
◆ perform the following calculations related to electrical/mechanical machines
   work done
   energy used/dissipated
   power developed/consumed
   efficiency in terms of input and output power/energy
◆ state the losses which occur in electrical machines

Part 1

Efficiency

In Chapter 7 the efficiency of transformers was covered. In this chapter we are mainly concerned with the efficiency of rotating machines, i.e. electric motors.

Power loss and efficiency of motors

It is not only low power factors that have to be taken into account when calculating the current demand of a motor.

No motor is perfect (i.e. 100% efficient) because power losses occur when the motor is transforming electrical energy into mechanical energy.

The power losses which occur are:
- mechanical in origin
- electrical in origin

Power loss which has a mechanical origin arises from one of two sources, namely
- frictional power loss
- windage power loss

Frictional losses are due to bearing friction and brush friction on certain motors (for example universal). Windage loss (ventilation loss) is due to the effort needed to circulate the ventilation (wind) to cool the motor.

Power loss which is electrical in origin occurs
- in the stator windings and rotor conductors (termed I^2R loss) and
- in the iron circuit (stator and rotor cores)

The "I^2R loss" or "copper loss" occurs because of the current flow in the stator and rotor conductors.

The "iron loss" is due to a combination of two types of power loss
- eddy-current loss (covered in Chapter 1)
- hysteresis loss (also covered in Chapter 1). This is caused by the continually reversing magnetic fluxes within the machine.

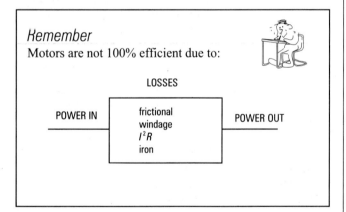

Remember

Motors are not 100% efficient due to:

LOSSES

POWER IN → [frictional / windage / I^2R / iron] → POWER OUT

Keeping power losses in electrical machines to a minimum

The maintenance electrician can ensure that:

Frictional losses

Bearings are periodically greased (except in the case of sealed-for-life bearings), but not overgreased, and grease which has congealed (solidified) is replaced.

Worn bearings are replaced.

The shaft and coupling of the motor are correctly aligned to the driven machine (pump, gearbox and so on).

On commutator and slipring motors, brushes are correctly tensioned and slide freely in their holders.

Worn brushes are replaced.

Windage (ventilation) losses

ventilation holes are clear and not blocked with dust

Note: Windage (ventilation) loss depends on the type of motor enclosure (i.e. a totally enclosed fan-cooled machine will have more windage loss than an open-type machine)

Iron losses

The electrical machine design engineer can ensure that:
– eddy currents are reduced by laminating the cores of stators and rotors
– hysteresis loss is reduced by choosing the most suitable core material

Copper losses

I^2R losses are reduced by incorporating larger c.s.a. copper windings, however, this may mean an increase in the machine size to accommodate the additional copper.

Calculation of efficiency

The efficiency of an electrical motor is the ratio of the electrical power output from the motor to its power input and is calculated by using:

$$\text{efficiency } (\eta) = \frac{\text{output power}}{\text{input power}} \times 100\%$$

Note: The greek letter η (termed eta) is the symbol for efficiency.

Example

Let's calculate the efficiency of a 230 V motor that takes a current of 15A and has an output power of 2880 watts.

$$(\eta) = \frac{\text{output power}}{\text{input power}} \times 100\%$$

$$= \frac{2880}{230 \times 15} \times 100\%$$

$$= 83.5\%$$

Example

Calculate the current taken by a 2 kW, 230 V single-phase motor working at full load with an efficiency of 70% and power factor of 0.6.

$$\eta = \frac{\text{output power}}{\text{input power}} \times 100\%$$

$$\therefore \text{ input power} = \frac{\text{output power}}{\eta} \times 100\%$$

$$= \frac{2000}{70} \times 100$$

$$= 28.57 \times 100$$

$$= 2857\text{W}$$

$$P = UI \cos \theta$$

$$\therefore \quad I = \frac{P}{U \times \cos \theta}$$

$$= \frac{2857}{230 \times 0.6}$$

$$= 20.7 \text{ A}$$

Try this

1. Calculate the full load current of a 10 kW, 230 V single-phase a.c. motor if the efficiency and power factor at full load are 80% and 0.8 lagging respectively.

2. The following results were recorded during a test on a single-phase motor:

Mechanical output = 1.2 kW
Supply volts = 240 V
Line current = 15 A
Power input = 1560 W

Calculate:
  (a) the efficiency, and
  (b) the power factor of the motor.

Efficiency calculations (three-phase motors)

Example
Calculate the full load current of a 40 kW, 400 V three-phase a.c. motor if the efficiency and power factor at full load are 85% and 0.85 lagging respectively.

$$\eta = \frac{\text{output power}}{\text{input power}} \times 100\%$$

$$\therefore \text{ input power} = \frac{\text{output power}}{\eta} \times 100\%$$

$$= \frac{40}{85} \times 100$$

$$= 47 \text{ kW}$$

Remember from Chapter 5,

$$\text{three-phase power} = \sqrt{3} U_L I_L \cos \theta$$

$$\therefore \quad I_L = \frac{\text{three phase power}}{\sqrt{3} U_L \cos \theta}$$

$$= \frac{47 \times 10^3}{1.732 \times 400 \times 0.85}$$

$$= 79.8 \text{ A}$$

Try this

The following results were recorded during a test on a three-phase motor:

Mechanical output = 50 kW
Supply volts = 400 V
Line current = 98 A
Power input = 58 kW

Calculate:
  (a) the efficiency and
  (b) the power factor of the motor

Work, energy and power

Let's first look at the physical quantities in Table 9.1 before considering work, energy and power, so we can clearly see their relevance in subsequent calculations on these topics.

Table 9.1

Physical (Symbol) Quantity	Unit (Symbol)	Definition
Force (F)	newton (N)	Force is a dynamic influence on a body or object which can cause a change in either the shape or the motion of the body or object.
Mass (m)	kilogram (kg)	Mass is the amount of matter which is contained in a body or object.
Weight	newton (N)	Weight is the downward force due to the mass of a body or object.

There is a connection between the mass of an object and its weight which is:

Weight = mass × 9.81 N

Note: The 9.81 Newtons is the force required to raise a mass or load of 1 kg against the effect of gravity.

For approximate calculations 10 can be used.

Example
Calculate the weight of a bundle of steel conduit which has a mass of 100 kg.

Weight = mass × 9.81 N
= 100 × 9.81 N
= 981 N

Try this
Determine the weight of a transformer with a mass of 400 kg.

Work
Electrical motors, as with any other type of motors, have to do work and this must be taken into account when determining the size of motor required for a given task.

Work is done whenever an object is moved by applying a force to it, and is calculated from:

Work done = force applied × distance moved

$$W = F \times d$$

where

force is in Newtons and
distance is in metres.

The unit of work is the joule (J) which is defined as the amount of work done when a force of 1 Newton acts through a distance of 1 metre.

1 J = 1 Nm

Example
The work done in lifting a mass of 10 kg through a height of 6 m is

Weight = mass × 9.81 N
= 10 × 9.81 N   = 98.1 N
W = $F \times d$
= 98.1 N × 6 m
= 588.6 Newton metres (Nm)
OR 588.6 joules (J)

Try this
A hoist lifts a mass of 500 kg through a vertical distance of 30 m to the top of a high rise building. Calculate the work done by the hoist in kilojoules.

Energy

Energy is the ability to do work. There are many forms of energy, for example chemical, thermal, mechanical and electrical. The unit of all forms of energy is also the joule (J), however it is often more convenient to measure electrical energy in kilowatt-hours (kWh).

$$
\begin{aligned}
3{,}600{,}000 \text{ J} &= 1 \text{ kWh} \\
\text{from} \qquad J &= W \times t \\
&= 1000 \text{ W} \times 60 \text{ min} \times 60 \text{ sec} \\
&= 3\,600\,000 \text{ J}
\end{aligned}
$$

Note: Energy and work are interchangeable since energy must be used to do work.

Both are measured in newton metres or joules.

So:

Energy used (W) = Work done (W) = Force applied (N) × Distance moved (m)

Example

Calculate the energy required to raise a mass of 6 kg through a vertical distance of 12 m.

$$
\begin{aligned}
\text{Work done} &= \text{force} \times \text{distance} \\
&= \text{mass} \times 9.81 \times \text{distance} \\
&= 6 \times 9.81 \times 12 \\
&= 706.32 \text{ Nm} \\
&\text{or } 706.32 \text{ J}
\end{aligned}
$$

Try this

An item of switchgear has a mass of 150 kg.

Calculate:
(a) the weight of the switchgear
(b) the energy used in raising the switchgear a distance of 3 m above finished floor level.

Example

A 250 V d.c. generator provides a current of 10 A. Calculate the energy dissipated in 2 minutes.

$$
\begin{aligned}
P &= U \times I \\
&= 250 \times 10 \\
&= 2500 \text{ Watts}
\end{aligned}
$$

$$
\begin{aligned}
\text{Energy} &= P \times t \\
&= 2500 \times 120 \\
&= 300\,000 \text{ Joules (or 300 kJ)}
\end{aligned}
$$

Try this

Calculate the energy dissipated in 1 hour by a 250 V, 20 A d.c. generator. (Answer in megajoules)

The efficiency of a motor can also be expressed as the ratio between the useful energy output and the total energy input, thus

$$\text{Efficiency } (\eta) = \frac{\text{useful energy output}}{\text{total energy input}} \times 100\%$$

Example

Calculate the energy in kWh required to lift a mass of 250 kg through 200 metres if the efficiency of the hoist is 30%.

Useful work done = force × distance

$$= 250 \times 9.81 \text{ N} \times 200 \text{ m}$$

$$= 490500 \text{ Nm (joules)}$$

Transpose the formula for efficiency above:

$$\text{Total work done} = \frac{\text{useful work done}}{\eta} \times 100$$

$$= \frac{490500}{30} \times 100$$

$$= 1635000 \text{ joules}$$

As the energy is required in kWh :

$$= \frac{1635000}{3600000}$$

$$= 0.454 \text{ kWh}$$

Power

Power (P) is a measure of how quickly work is done.

Thus P (in watts) = Time rate of doing work

$$= \frac{\text{work done (in joules)}}{\text{time taken (in seconds)}}$$

or watts $= \dfrac{\text{joules}}{\text{time}}$

$$W = \frac{J}{t}$$

One watt is equivalent to work being done at the rate of one joule per second (1 W = 1 J/s)

Also since energy used = work done

$$\therefore \text{ power (in watts)} = \frac{\text{energy (in joules)}}{\text{time (in seconds)}}$$

Transposed:

$$\text{Energy} = \text{power} \times \text{time}$$

Example

Calculate the power rating of an electric hoist motor required to raise a load of 200 kg at a velocity of 4 m/second (ignore efficiency).

$$\begin{aligned} \text{Weight} &= 200 \times 9.81 \text{ N} \\ &= 1962 \text{ N} \\ W &= f \times d \\ &= 1962 \times 4 \\ &= 7848 \text{ Nm or } 7848 \text{ J} \end{aligned}$$

$$\begin{aligned} P &= \frac{W}{t} \\ &= \frac{7848}{1} \qquad = 7848 \text{ J/s} \quad = 7848 \text{ watts} \end{aligned}$$

Try this

An electric motor drives a pump which lifts 1000 litres of water each minute to a tank 25 m above the main storage tank. Ignoring efficiency, calculate the power rating of the motor.

Note: 1 litre of water weighs 9.81 N.

Efficiency, work, energy and power

Power losses which occur in electrical machines are either _____ or _____ in origin.

Bearing _____ causes a _____ power loss.

The cooling system of an electric motor will cause a _____ power loss.

The I^2R loss in the stator windings of a motor is also known as the _____ _____ .

Iron loss is due to (1) _____ loss and (2) _____ loss.

The efficiency of an electrical machine is given by the _____ of _____ power to the _____ power and is expressed in _____.

The unit of force and also weight is the _____.

The mass of one kilogram will exert a force of _____ when lifted against the effects of _____.

The unit of work is the _____.

Energy is the ability to do _____ and the unit of all forms of energy is the _____.

Power is a measure of how quickly _____ is _____, and its unit is the _____.

The efficiency of a motor can be expressed as the ratio between the _____ energy _____ and the total energy _____.

Part 2

Practical problems involving efficiency

A single-phase motor drives a pump which raises 600 litres of water per minute to the top of a building 10 m high. Calculate the power that the motor must provide if the pump is 50% efficient.

$$\text{Weight} = 600 \times 9.81 \text{ N}$$
$$= 5886 \text{ N}$$

$$W = F \times d$$
$$= 5886 \text{ N} \times 10 \text{ m}$$
$$= 58860 \text{ Nm or } 58860 \text{ J}$$

$$\text{Pump output power} = \frac{58860}{60}$$

$$= 981 \text{ watts}$$

$$\text{Pump input power} = \frac{\text{pump output power}}{\eta} \times 100$$

$$= \frac{981}{50} \times 100$$

$$= 1962 \text{ watts}$$

$$\text{Motor output power} = \text{pump input power}$$

$$\therefore \text{Motor output power} = 1962 \text{ watts}$$

Try this

Calculate the output power rating of an electric hoist motor which is required to raise a load of 250 kg at a velocity of 2 m/second. The efficiency of the hoist is 40%.

Points to remember

Work, power and efficiency calculations

Work done   = _____ × _____

Weight   = mass × _____

Power   = $\dfrac{\text{(in _____)}}{\text{(in _____)}}$

The_____ power of the motor will be the
_____ as the _____ power of the pump
in Figure 9.2 below.

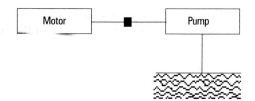

Figure 9.2

One_____ of water weighs one kilogram.

The force required to raise 1 litre of water against the effect of
gravity is _____.

1000 Nm is the same as 1000_____ when
calculating work done.

Self assessment short answer questions

1. State the units which EACH of the following are
   measured in:
   (a) power
   (b) energy
   (c) work

2. Define EACH of the following terms:
   (a) work
   (b) energy
   (c) efficiency

3. (a) Determine the weight of an electric motor with a
       mass of 250 kg.
   (b) Calculate the work to be done in lifting the motor
       from its bedplate to a height of 2 m ready for loading
       it onto a trailer.

4. Calculate the energy required to raise a mass of 10 kg through a vertical distance of 15 m.

6. (a) State two losses that occur in a.c. induction motors.
   (b) An electric motor has an efficiency of 80% and output of 5 kW. Calculate the motor input.

5. Determine the power required to lift a mass of 300 kg through a height of 10 m in 40 seconds.

7. Calculate the energy in kWh required to lift a mass of 200 kg through 200 metres if the efficiency of the hoist is 40%.

8. Calculate the output power of a 200 V d.c. motor which draws a current 10 A. The efficiency of the motor is 80%.

10

Instruments

Answer the following questions to remind yourself of what was covered in Chapter 9.

1. Explain why the output energy of an electrical machine is less than the input energy.

2. The work done in raising a load through a height of 5 m is 1500 J. Find the required raising force.

3. The efficiency of an electric motor is 75%. Calculate the energy input when the energy output is 150 kJ.

4. If the energy required to raise a mass through a vertical distance of 10 m is 600 Nm, calculate the mass.

Part 1

Measuring instruments

As electrical quantities have no obvious physical properties, the only way to obtain a reliable indication of their presence and magnitude is by measuring the effects that they produce.

Electrical measuring instruments use a current, or the presence of an electrical charge, to create a physical change which can be translated into some form of visible output.

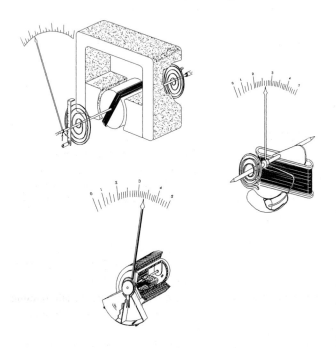

Figure 10.1   Measuring instruments

Measuring instruments have to be accurate and reliable if they are to be of any use. The accuracy and reliability of an instrument will depend on the soundness of its design, its construction and the choice of components used in its construction.

It is not the purpose of this chapter to consider the construction of such instruments but some attention will be given to their operating principles so that you can have a better understanding of the techniques involved in their application.

Analogue instruments

In the majority of cases an analogue instrument takes an electric current and uses this to produce a mechanical deflection. The deflecting force can be the force on a conductor in a magnetic field, as in the case of the moving coil instrument, or the force exerted between the poles of a magnet as in the moving iron. In either case, an electromagnetic device is used to create the deflecting force. The force will depend on the current flowing in the instrument. The greater the current – the greater the force.

In order to measure the strength of the deflection, it must be restrained by a controlling force. The controlling force is usually provided by two spiral springs which hold the movement at rest when the amount of torque exerted by the deflecting force is matched with the controlling torque of the springs. When there is no current in the measuring instrument the control springs will hold the movement at zero.

In addition to deflection and controlling forces, it is necessary for some form of damping to be provided for analogue instruments. Without a damping mechanism, a meter movement will oscillate to and fro for a considerable time before coming to rest at its true deflection. This makes it very difficult to obtain an accurate reading, more especially so if the deflecting current changes before the movement has come to rest.

There are various types of analogue measuring instruments. The two basic types, as previously mentioned, are the moving coil and moving iron.

Moving coil

A coil of wire suspended between the poles of an electromagnet produces a force of deflection which is controlled by hair springs. The direction of the deflecting force depends on the direction of current flow in the coil therefore this instrument cannot respond to an alternating current. This type of instrument is primarily intended for use in d.c. circuits only. With the aid of rectification, Figure 10.3, the moving coil instrument can be used for a.c. measurement provided that certain modifications are carried out. The damping technique used in moving coil instruments is usually eddy current damping. This is achieved by winding the deflection coil on an aluminium former which acts as a damper. When the former is in motion in the magnetic field of the instrument a current is induced which sets up a magnetic flux and consequently a force to oppose the movement of the coil.

The scale of the moving coil instrument is linear i.e. the scale divisions are the same size from zero to full scale deflection (f.s.d.).

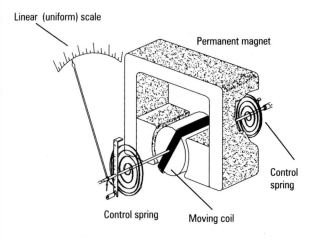

Figure 10.2   Moving coil meter

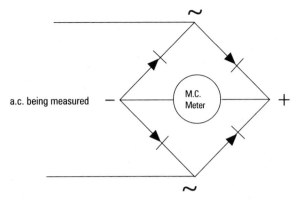

Figure 10.3   Simplified diagram of a rectified moving coil
meter

Moving Iron

Moving iron instruments work on the principle of magnetic attraction or repulsion depending on their design. They normally use a coil spring controlling mechanism but some may use a simple gravity device. The moving iron instrument is deflected by direct current or r.m.s. alternating current therefore it does not need a rectifier when connected to a.c. circuits.

Without damping, a moving iron instrument swings quite appreciably before coming to rest in its deflected state. The damping mechanism most commonly used is in the form of an air vane or damper which is enclosed in a chamber. The air pressure built up by the damper acts as a decelerating force in either direction and quickly brings the movement to rest. The scale of a moving iron instrument is non-linear (uneven) being cramped at the beginning because the torque or turning power depends upon the square of the current flowing.

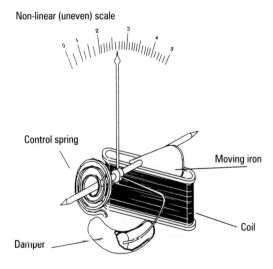

Figure 10.4   Attraction type moving iron meter

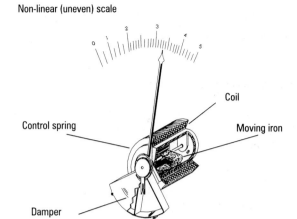

Figure 10.5   Repulsion type moving iron meter

Electrodynamic

This type of moving coil instrument uses an electromagnet, instead of a permanent magnet, to produce the magnetic field which surrounds the coil. Although electrodynamic ammeters and voltmeters can be found, the main application for this type of instrument is as a wattmeter.

The control of the movement is by hairsprings. As an electrodynamic or dynamometer wattmeter the deflection is produced between the fixed current coil and the moving voltage coil acting at the same instant giving a scale reading which is proportional to the true power in the measured circuit.

Air damping is usually provided for this type of instrument and the scale is linear.

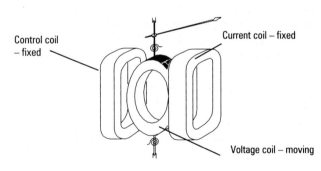

Figure 10.6   Electrodynamic meter

Advantages and disadvantages of moving coil and moving iron instruments

Moving coil instrument

Advantages

     High sensitivity
     Linear (uniform) scale
     Not seriously affected by external magnetic fields
     Can be used in any plane
     Can be adapted for centre-zero operation
     Can easily be used as a multi-range instrument

Disadvantages

     Expensive compared to moving iron
     Delicate construction
     Will operate only on d.c. unless used with a rectifier

Moving iron instrument

Advantages

     Very robust instrument
     Cheaper than moving coil
     Can withstand heavy overloads
     May be used on a.c. or d.c. circuits

Disadvantages

     Not as accurate as a moving coil type
     Non linear (uneven) scale
     Can be affected by external magnetic fields
     Can only be used on other ranges by changing the coil

Parallax error

With analogue instruments it is possible to misread the pointer position on the scale if the eye is not positioned exactly above the pointer (i.e. at 90° to the meter scale). This is a case of "parallax error", and most analogue meters have an anti-parallax mirror placed behind the pointer so that the effect is minimised (Figure 10.7).

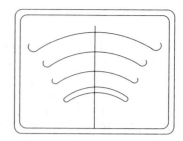

Correct mirror view
Positioned at 90° to the meter scale, the reflection of the pointer is hidden by the pointer itself.

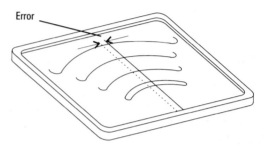

Viewed from left
Viewed from the left the apparent reading is high.

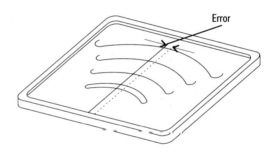

Figure 10.7   *Viewed from right*
Viewed from the right the apparent reading is low.

Digital instruments

Digital instruments fulfil the same function as analogue instruments in that they are used to take measurements and to convey the information to the user.

The operating principle is, however, quite different to the analogue types previously discussed.

The most obvious difference is the display. This is generally an alpha-numeric arrangement with seven elements (segments) although more sophisticated types are readily available (Figure 10.8).

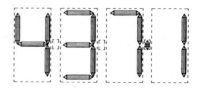

Figure 10.8   *Digital display*

Another difference is that the digital instrument requires a power source before it can be operated. This will be a battery or batteries, in the case of the portable instrument, or a power supply in the case of bench or panel instruments.

The operation of the digital instrument is the biggest fundamental difference. The measured signal will be a voltage, or in the case of current and resistance measurement, the voltage drop across a resistor of known value.

Advantages
> Robust
> No moving parts
> Accurate
> Clear (easy to read) display

Disadvantages
> Requires a battery or a power supply

Note: The digital display could be a liquid crystal display (L.C.D.) or the type which employs Light Emitting Diodes (L.E.D.) one for each segment of the display.

Remember
Analogue and digital instruments have their advantages and disadvantages and the choice of the "best" instrument depends on the application you have in mind for it. As a rough guide to the features of the instruments the following points are useful:

- An analogue instrument does not require a battery or power supply, whereas a digital instrument does.
- A digital instrument is generally more accurate than an analogue instrument.
- Both types are portable.
- Analogue instruments require a deflecting force, a controlling force and a damping force for the smooth operation of the pointer movement, these are mechanical disadvantages.
- A digital display is usually clearer to read than an analogue type.

Ammeters and voltmeters

When considering ammeters and voltmeters, one can say that there is no fundamental difference between them.

In the case of the ammeter, the basic movement is designed to give full scale deflection at a very low current, far lower than would be considered suitable for practical applications. To extend the range so that the instrument can be of some practical use, a shunt resistor is connected in parallel with the meter movement (Figure 10.9).

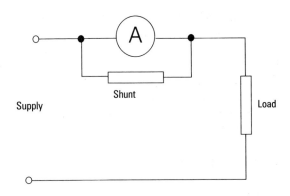

Figure 10.9   *Ammeter shunt*
          Note: *Shunts are "low ohmic" value resistors.*

Voltmeters

The same meter movement is used for voltage measurement but in the case of the voltmeter, the range is extended by connecting additional resistance in series with the meter movement. This series resistance is known as a multiplier. (Figure 10.10)

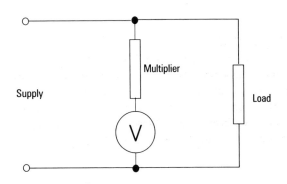

Figure 10.10   *Voltmeter multiplier*
          Note: *Multipliers are "high ohmic" value resistors.*

Remember
The "shunt" extends the range of a moving coil ammeter to read values of current higher than the instruments movement is designed for and similarly the "multiplier" is used to extend the range of a moving coil voltmeter.

Measuring instruments

The two basic types of analogue instrument are:

_____ and _____.

A moving coil meter is fitted with a _____ to enable it to measure a.c.

A moving iron instrument has a _____ scale whereas a moving coil instrument has a _____ scale.

A _____ instrument requires a battery or a _____ supply for it to operate.

Shunts are used to _____ the range of moving coil _____, and multipliers are used to _____ the range of moving coil _____.

Analogue instruments require a _____ force, a _____ force and a _____ force to enable the pointer to move smoothly across the meter scale.

_____ meters are fitted with an anti-parallax mirror to keep _____ _____ to a minimum.

Part 2

Monitoring and metering

It is not enough to know that a circuit works, it is also necessary to be able to check that it is working in the way that it was designed. In many cases where a circuit supplies a particular piece of equipment the performance of the equipment confirms to some extent the suitability of the circuit. Even so it is often necessary to take measurements of voltage, current and power to check that the equipment is working to its specification.

Analogue voltmeter

Digital clamp-on ammeter

Figure 10.11

Voltage measurements

The fact that voltage readings must always be taken on live circuits means that every safety precaution should be taken. Barriers and screens may be necessary to offer protection from touching exposed live conductors. Wherever possible voltages should be taken where all live conductors are insulated so that only the probe tips of the test equipment can touch the live conductors.

In many cases it is possible to connect the voltmeter when the supply is switched off and then the readings can be taken from a safe distance when the circuit is re-energised. Test probes should conform with Health and Safety Executive Guidance Note GS38, as the example shown in Figure 10.12.

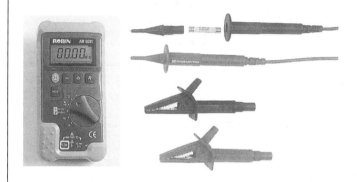

Figure 10.12   Fused test prods
Reproduced with kind permission from Robin Instruments.

Voltmeters come in many different forms. They can be meters designed for one voltage range, say 0 200 V. In this case they cannot be used for voltages above that and voltages below 50 V would be inaccurate. Some of the digital meters are completely automatic and will select the range most suitable for the voltage being measured. Regardless of whether it is a single range or multi-range instrument, analogue or digital, there are certain factors that must always be considered when measuring voltages:

* the minimum and maximum readings required
* the nature of the voltage i.e. a.c. or d.c.
* safety

Let us now consider the actual use of an instrument.

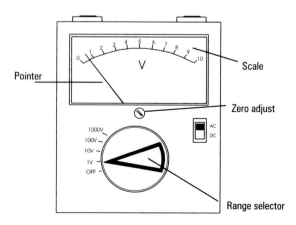

Figure 10.13

The instrument shown in Figure 10.13 is an analogue multi-range voltmeter with a single scale. Although the scale is numbered from 0 to 10 this does not mean it will always be the 0 to 10 volts that are indicated. If the selector switch is pointed at 10 V then that is the maximum the instrument will indicate. When the selector switch is at 1 V then the 10 V on the scale will be equivalent to 1 V and if the pointer indicates 5 this will in fact be 0.5 V. Similarly if the range switch is pointing at 100 V then the 10 V on the scale should be read as 100 V and the other calibrations should all have a zero added on to them, so if the pointer is on 6 this indicates 60 V. With the selector switch on 1000 V then 2 zeros are added.

Safety

Voltage readings are taken on **LIVE** equipment.

Every safety precaution must be used.

Try this
Voltage readings

Write the reading of each meter in the space provided.

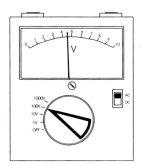

Figure 10.14   Reading:_____

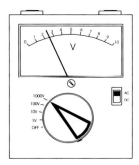

Figure 10.15   Reading:_____

Voltage readings should always be taken across a difference in potential. For example across the supply.

In Chapter 5 the voltage arrangements for star connected supplies was explained. Although a supply voltage, phase to neutral, is 230 V if a star connected three phase supply is used the maximum voltage available is 400 V. This means that if there is any doubt as to what the voltage may be the meter must be switched to read the maximum and then switched back if necessary.

The supply company will state what their nominal voltage will be at the supply intake. They have a legal obligation to keep that within + 10%, –6% of what they state. Within the building any drop in voltage due to long cable runs and large loads is up to the consumer and their consultants. Generally the voltage drop within an installation is kept to no more than 4% of the supply voltage. To ensure this has not been exceeded voltage readings sometimes have to be taken at the load and these compared with the supply voltage. Excessive voltage drop can cause the malfunction of equipment and heat being produced in cables.

Note: Domestic, commercial and industrial voltage values were changed from 415/240 Volts, ±6%, to 400/230 Volts + 10%, –6% on 1 January 1995. Voltage tolerance levels will be further adjusted to ±10% by 2003.

Measuring high voltages

It is sometimes required to measure voltages that are up in the thousands of volts. An example of this is on the supply transmission and distribution systems. So that voltages can be monitored all the time voltmeters are built into control panels. To reduce the risk of danger and keep the voltmeter a practical size transformers are used on the high voltage cables.

Using this method the maximum voltage on the meter may be only about 50 V even though the scale is calibrated to read 11000 V.

Remember V.T.s were covered in more detail in Chapter 7.

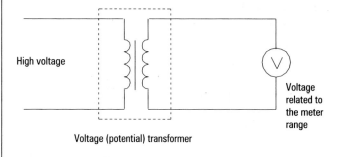

Voltage (potential) transformer

Figure 10.16

Current measurement

As current flows through a circuit to measure it the ammeter (flow meter) must be connected into the circuit. This means that the circuit has to be broken and the meter connected into it.

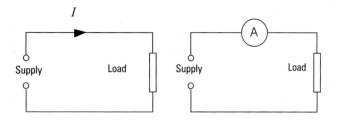

Figure 10.17

This can be dangerous as the connection leads and the meter have all got to be capable of carrying the maximum currents. These should never be connected when the circuit is energised.

Where current is to be monitored continually with panel mounted meters these meters may be wired into the circuit. It is more likely, especially if the currents are high, that current transformers would be used. A current transformer uses the magnetic field set up around a conductor carrying alternating current. The conductor acts as the primary and a secondary coil is placed around it. Remember C.T.s were also covered in more detail in Chapter 7.

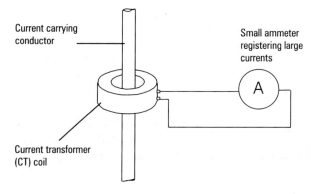

Figure 10.18

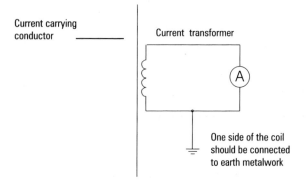

Figure 10.19

This method of monitoring current means that the current carrying conductor does not have to be broken as it passes through the current transformer coil. Only a single conductor should be used because if a twin cable with a phase and return was put through the coil the ammeter would read zero. This is because the magnetic field set up by the phase conductor would be cancelled out by the opposing field in the neutral conductor.

As current transformers are basically double wound transformers, but in this case the primary winding has only one turn, the same calculations can be used. Assuming a conductor carries a maximum load of 500 A and the ammeter is calibrated to show this when in fact only 5 A is flowing through the meter, the coil must have 100 turns

$$\frac{N_S}{N_P} = \frac{I_P}{I_S}$$

$$\frac{N_S}{1} = \frac{500}{5}$$

$$= 100 \text{ turns}$$

where
N_S = number of secondary turns

N_P = number of primary turns   = 1

I_P = Primary current   = 500 A

I_S = Secondary current   = 5 A

Warning

The CT coil must never be open circuited while the primary current is still flowing as dangerously high voltages may be induced in the windings of the coil. The coil must always be shorted out before the ammeter is disconnected.

Clamp-on ammeter (tong tester)

The current transformer, which is fixed as shown in Figure 10.18, consists of a complete coil which has the load conductor passing through it. To change this from one conductor to another means the loads and supplies must be switched off, the coil disconnected and then reconnected on the new load and then it can all be switched back on again. This takes time and can be very inconvenient.

However, the theory using the current transformer has been adapted so that a portable easy to operate meter has been developed.

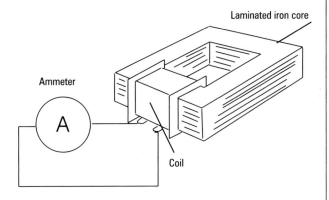

Figure 10.20

As with most double wound transformers the coil is wound on a former and fitted on a laminated core. This iron core is different to others insomuch as it is made to open up and allow room for a conductor to go into the centre.

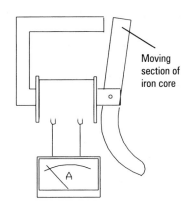

Figure 10.21

The moving section of laminated iron core is sprung so that when the gap is closed there is a tight joint between the two surfaces. The method of opening the core and where it opens depends on particular manufacturers. An example of a complete "clamp on" ammeter is shown in Figure 10.22.

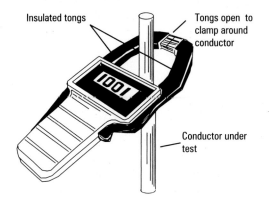

Figure 10.22   Digital clamp on ammeter

This same method for monitoring the current can be used when measuring power and power factor. Some of the meters using this method have automatic range switching, others have an external switch. It is not uncommon for this type of meter to be capable of measuring up to 1000 A.

> ### Remember
> The clamp-on ammeter allows current measurements to be taken without having to disconnect the supply and the circuit.

Measurement of power

In a simple d.c. circuit, the power can be measured by multiplying the readings obtained from an ammeter and a voltmeter.

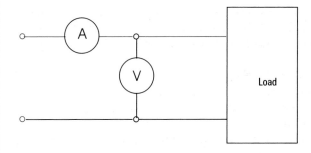

Figure 10.23

$$P = U \times I$$

In an a.c. circuit, this is no longer reliable since, as we have seen earlier in this book, the current and voltage may no longer be in phase with each other and the result obtained will be the VoltAmps rather than the true power of the circuit.

The electrodynamic (dynamometer) wattmeter connected as shown in Figure 10.24 will read the product of current and voltage at the same instant and will indicate the true power of the circuit.

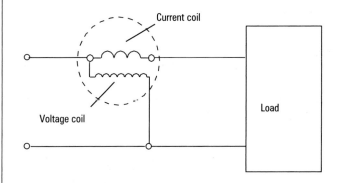

Figure 10.24   Circuit diagram of wattmeter

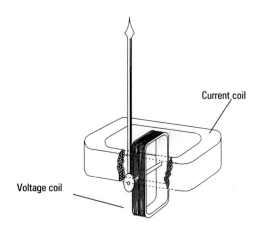

Figure 10.25   Dynamometer wattmeter

Try this

A wattmeter indicates 0.46 on a scale of 0 to 1 when the multiplying factor is 250.

(a) What is the power indicated by the instrument?

(b) If the current and voltage ranges are changed so that the factor becomes 150 what should the scale deflection read?

As a portable instrument, the wattmeter has to be suitable for a wide variety of currents and voltages. If, for example, a wattmeter has three voltage and four current ranges this means that the deflection will have twelve possible interpretations.

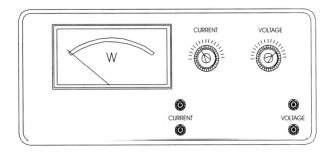

Figure 10.26   *A practical meter with voltage and current connections*

A scale of such complexity is obviously impractical so a single scale will be used and this will be read and multiplied by a factor determined by the current and voltage ranges selected.

Example

A wattmeter reads 0.7 on a scale of 0 to 1 when the current range is 0–1 A and the voltage range is 0–100 V. If the multiplying factor accompanying this choice of ranges is 100 then the power indicated is

$$0.7 \times 100 = 70 \text{ Watts}$$

If however the instrument indicated 0.35 on the 0–1 A and 0–200 V ranges where the multiplying factor is 200 then the power indicated would be

$$0.35 \times 200 = 70 \text{ Watts}.$$

Wattmeter connections

As power is the product of voltage and current it follows that a wattmeter is basically a voltmeter and ammeter combined. Providing the voltage coil is connected as a voltmeter across the potential, and the current coil is connected as an ammeter in series, there should be no problems using these. However, should the windings be incorrectly connected serious damage can be caused to the meter and the circuit.

There are usually 4 terminals on a wattmeter, 2 for the voltage coil and 2 for the current.

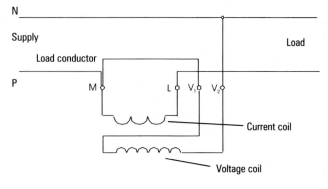

Figure 10.27   *Meter connections*

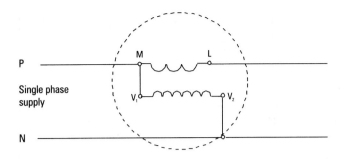

P

Single phase
supply

N

Figure 10.28   Circuit diagram

Measuring power in three phase circuits

Where the power is to be measured in three phase circuits this can become more complex. If, however, the loading on each phase is the same this can be carried out using just one wattmeter.

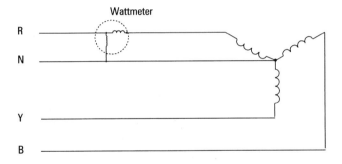

Figure 10.29

The wattmeter is connected between one phase and the star point as though it is a single phase supply. The reading on the wattmeter can now be multiplied by three to find the total power of the load. This method can only be used on balanced three phase loads where the star point is available.

The reading obtained by a wattmeter is the true power in the circuit and if this is at all reactive the power will not be equal to the voltage times the current. By taking voltage, current and power readings of the same load its power factor can be calculated from

$$\text{power factor} = \frac{\text{true power (wattmeter reading)}}{\text{voltage} \times \text{current}}$$

The answer should always be less than one.

Remember this was also covered in Chapter 4.

Note: This is a simple but cumbersome method of measuring power factor and it gives no indication whether the power factor is lagging or leading like a purpose made power factor instrument does (see Figure 4.36).

Try this

Study a wattmeter.

1.   Sketch the terminal arrange-
     ments

2.   Write down how the different
     ranges are set.

3.   State any precautions that are shown on the meter.

Measuring voltage, current and power on high voltage/current a.c. systems

Figure 10.30 shows the use of instrument transformers for metering purposes on high voltage and high current a.c. systems.

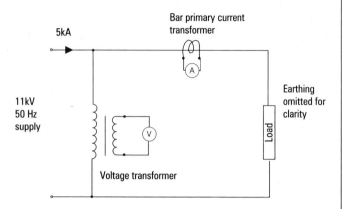

Figure 10.30

Figure 10.31 shows the connection of a single-phase wattmeter used to measure the power of a high voltage and high current a.c. system.

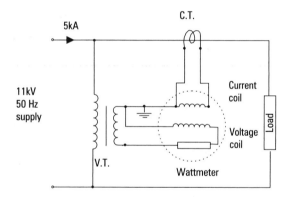

Figure 10.31

The current and voltage transformer "isolate" the wattmeter from the high current/voltage system.

Energy meters

Energy is the measurement of power taken over a period of time.

It is this that the electricity companies use to monitor the electricity used by a consumer. The meter is often referred to as the kilowatt-hour meter as these are the quantities that are measured.

Energy is measured in joules.

One joule is equivalent to 1 watt used in one second.

This value (the joule) is too small for the supply companies to use as the values of the energy used by each consumer would be up in the millions. For example one unit of electricity (1 kilowatt-hour) is equal to 3600000 joules. This can be calculated from

$$1 \text{kW} \qquad = 1000 \text{ watts}$$

$$1 \text{ hour} = 60 \text{ sec.} \times 60 \text{ min.} \quad = 3600 \text{ sec.}$$

$$1000 \times 3600 \qquad = 3600000 \text{ joules}$$

The energy, or kilowatt-hour meter, is connected in a similar way to the wattmeter for it also has current and voltage coils.

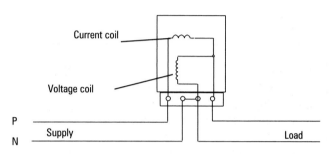

Figure 10.32

On the older type meter the voltage and current coils are arranged in such a way that the magnetic fields being produced by them induce a current into an aluminium disc and make it act as a motor. The revolutions of the disc directly relate to the amount of energy consumed. The recently developed meters use electronic pulses to measure the energy used. The arrangements of the connections are the same.

Connection of three phase energy meters

In some installations the measurement of the energy used on three phase supplies is the same as three single phase supplies.

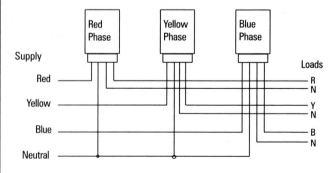

Figure 10.33

Care would have to be taken to keep the loads across the three phases roughly in balance.

It is more usual on large installations to have a three phase or "poly phase" meter installed (Figure 10.34).

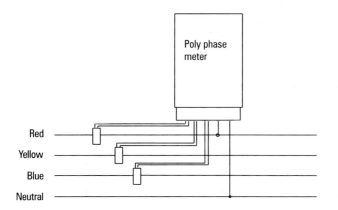

Poly phase meter

Red

Yellow

Blue

Neutral

Figure 10.34

Remember

There are many different reasons for monitoring the voltage, current, power and energy. The correct use of the appropriate meters is important not only for getting accurate results but also for safety reasons. The clamp on ammeter now makes current measurement more practical but only if it is used correctly. The measurement of energy is often limited to the supply companies for accounting purposes.

Where there are large loads in use the current is monitored using current transformers around the main supply conductors.

The current transformers are often installed on the main supply cables to a busbar chamber.

Try this

Look at the intake of a domestic consumer's premises.
Draw a block diagram showing the supply company's equipment.
Label the parts and note the type of energy meter.
Find out the tariff that the consumer's electricity is costed on.

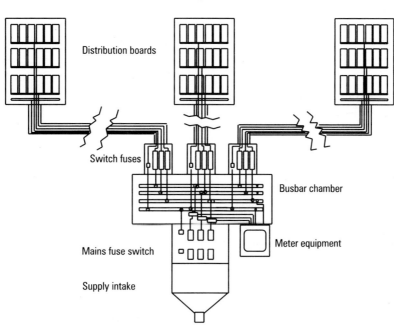

Distribution boards

Switch fuses

Busbar chamber

Meter equipment

Mains fuse switch

Supply intake

Figure 10.35

Monitoring and metering

_____ _____ should conform with Health and Safety Executive Guidance Note GS38. A control panel voltmeter is connected to a _____ _____ to enable it to measure high voltages, and a control panel ammeter is connected to a _____ _____ to enable it to measure _____ currents.

A clamp-on ammeter allows you to measure the _____ taken by a load without having to _____ the supply and the _____.

A dynamometer wattmeter has fixed _____ coils and a moving _____ coil.

The wattmeter measures the _____ power of the circuit and the voltmeter and ammeter measure the _____ power of the circuit.

It is possible to measure the total power of a _____ three-phase load using only _____ _____.

Energy meters measure the amount of _____ taken over a period of _____ in _____.

An energy meter has a _____ coil connected in series with the load and a _____ coil connected in parallel with the load.

Short answer self assessment questions

1. State two advantages and one disadvantage of a moving coil instrument.

2. (a) Name the two different types of moving iron meter.
   (b) What type of scale doe these meters have?

3. (a) With the aid of a diagram, show how a moving coil instrument may have its range extended to measure 0–10 A d.c.
   (b) Explain the purpose of a clamp-on ammeter.

4. Draw a circuit diagram showing how each of the following can be measured in a single-phase a.c. circuit supplying a load.
   (i) voltage, (ii) current (iii) true power

5. Identify:
   (a) the type of meter shown in Figure 10.36.
   (b) the meter components labelled A and B.

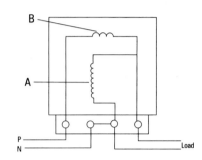

Figure 10.36

6. With the aid of a diagram show how the voltage and current may be measured in an a.c. circuit supplying a load at 11 kV and 1500 A.

7. Draw on Figure 10.37 below
   (a) the field around each current carrying conductor
   (b) the main field between the two poles of the permanent magnet. Also indicate the direction of the forces acting on each conductor (by applying Flemings L.H. Rule) and the direction the pointer will move.

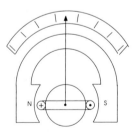

Figure 10.37

8. (a) State two advantages of using a digital instrument compared with an analogue instrument.
   (b) What is fitted to an analogue type meter to minimise parallax error?

End test – short answer

1. (a) Redraw the diagram shown in Figure ET.1, and show the resultant magnetic field on the diagram.
    (b) State whether the force between the magnets is one of attraction or repulsion.

Figure ET.1

2. Calculate the e.m.f. induced in a conductor of length 20 cm which is moving through a magnetic field of flux density 0.4 T at a velocity of 60 m/s.

3. Calculate the total capacitance of the circuit shown in Figure ET.2.

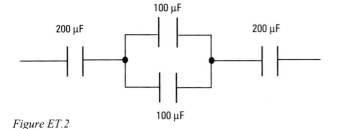

Figure ET.2

4. State:
    (a) two industrial applications of capacitors
    (b) the reason for connecting a discharge resistor to a large capacitor.

5. (a) Identify the type of motor represented in Figure ET.3.
    (b) State the purpose of the armature divertor.
    (c) How can the direction of rotation of this type of motor be reversed?

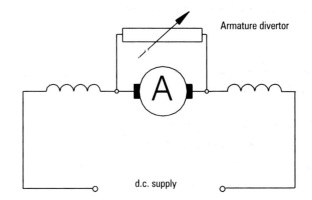

Figure ET.3

6. Calculate the terminal voltage of a d.c. generator which has an armature resistance of 0.3 Ω and a current flowing through it of 4 A when it generates an e.m.f. of 240 V.

7. For the circuit shown in Figure ET.4, calculate the
   (a) impedance
   (b) power factor

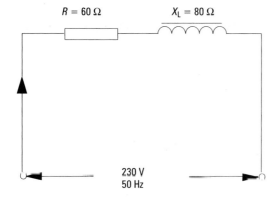

Figure ET.4

8. Draw a phasor diagram to scale, for the circuit shown in Figure ET.5, and from the phasor diagram determine the value of the supply current.

(10 A at a p.f. of 0.5 lagging)

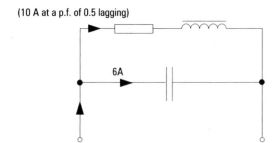

Figure ET.5

9. A small industrial premises has a three-phase load of 55 kW at 70 kVAr. Calculate
   (a) the kVA
   (b) the power factor

10. Draw a circuit diagram showing how a 12 V d.c. supply can be obtained from a 230 V, 50 Hz a.c. supply using a full-wave bridge rectifier.

11. (a) State the purpose of the slip-rings on a wound-rotor induction motor.
    (b) When are the slip-rings on this type of motor shorted out?
    (c) How can the direction of rotation of this type of motor be changed?

12. Determine:
    (a) the synchronous speed of a six pole a.c. induction motor when connected to a 50 Hz supply,
    (b) the speed of the rotor if the percentage slip is 5%.

13. A transformer with 1000 primary turns and 250 secondary turns is fed from a 230 V a.c. supply. Calculate:
    (a) the secondary voltage
    (b) the volts per turn

14. (a) Sketch:
        (i) the core arrangement of a single-phase bar-primary current transformer
        (ii) the winding arrangement for a single-phase voltage transformer (V.T.)
    (b) What is the purpose of a voltage transformer?

15. (a) Calculate the illuminance at point P on the working plane.
    (b) Will the level of illumination at point Q be greater or lower than that at point P?

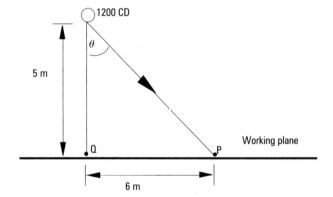

Figure ET.6

16. Redraw the circuit diagram shown in Figure ET.7 and include the missing components A, B and C.

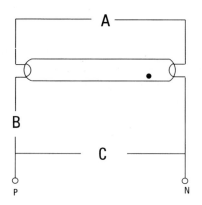

Figure ET.7

17. (a) Calculate the weight of an a.c. induction motor with a mass of 350 kg.
    (b) Determine the work to be done in lifting the motor from its bedplate to a height of 3 m ready for loading it onto a trailer.

18. (a) Electrical machines are not 100% efficient due to mechanical and electrical power losses. State ONE type of mechanical power loss, and ONE type of electrical power loss.
    (b) An electric motor has an efficiency of 75% and an output of 12 kW. Determine the motor input power.

19. (a) State the purpose for EACH of the following instrument components
    (i) multiplier
    (ii) shunt
    (b) What is fitted to a moving coil instrument to enable it to measure a.c.?

20. With the aid of a diagram show how the current and voltage may be measured in an a.c. circuit supplying a load at 1000 V and 500 A.

Multi-choice end test

Circle the correct answers in the grid on page 194.

1. Voltages of 400 kV are used for
   (a) generation
   (b) urban distribution
   (c) national transmission
   (d) factory distribution

2. A single phase motor is connected so that its voltage, current and wattage can be monitored.
   One set of readings gives
   $V = 240$ V, $I = 4.8$ A and $P = 920$ W.
   The power factor in this case is
   (a) 1.2
   (b) 18.4
   (c) 0.8
   (d) 0.5

3. Which of the following is correct for the bridge rectifier in the diagram?

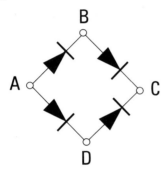

	A	B	C	D
(a)	dc–	a.c.	dc+	a.c.
(b)	dc–	dc+	a.c.	a.c.
(c)	a.c.	a.c.	dc+	dc–
(d)	a.c.	dc+	a.c.	dc–

4. The range selector switch on a voltmeter is set at 500 V and the scale is calibrated 0 to 10. If the needle points to 2.5 the actual voltage is
   (a) 500 V
   (b) 250 V
   (c) 125 V
   (d) 2.5 V

5. The current carrying conductor in the diagram will move

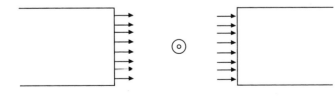

   (a) downwards
   (b) upwards
   (c) to the left
   (d) to the right

6. When applying Fleming's Left Hand Rule the second finger indicates the direction of the
   (a) motion
   (b) field
   (c) current
   (d) force

7.

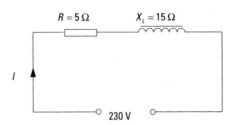

In the diagram the impedance of the circuit is
   (a) 4.47 Ω
   (b) 14.14 Ω
   (c) 20 Ω
   (d) 15.8 Ω

8.

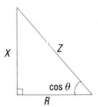

In the diagram if $X = 24$ and $R = 18$ the value of Z must be
   (a) 30
   (b) 42
   (c) 36
   (d) 24

9. Tungsten halogen tubular lamps should be mounted within
   (a) 10° of vertical
   (b) 60° of vertical
   (c) 60° of horizontal
   (d) 10° of horizontal

10. The power consumed by an inductor can be calculated from
   (a) $P = VI$

   (b) $P = VI \cos \theta$

   (c) $P = \dfrac{V}{I \cos \theta}$

   (d) $P = \dfrac{V}{I}$

11. The BS3535 shaver socket is an example of protection by
    (a) electrical separation
    (b) barriers and enclosures
    (c) Class 1 equipment
    (d) a solid earthed system

12. One disadvantage of a tungsten halogen lamp is that
    (a) it requires expensive control equipment
    (b) the quartz envelope runs at a high temperature
    (c) it is not suitable for display lighting
    (d) it has poor colour rendering

13. The waveforms in the diagram show that

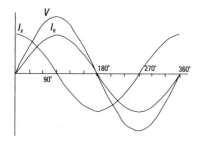

    (a) V is in phase with I_X
    (b) I_R is in phase with V
    (c) I_R leads I_X by 90°
    (d) I_X lags I_R by 90°

14. The capacitor connected across the supply to the circuit is to
    (a) control the current across the tube
    (b) limit the current through the choke
    (c) reduce the current in the supply conductors
    (d) smooth the current on starting

15. The predominant colour given from sodium lamps is
    (a) red
    (b) green
    (c) yellow
    (d) blue

16. The power factor of an a.c. circuit is found by
    (a) $\sin \theta = \dfrac{kW}{kVA}$
    (b) $\cos \theta = \dfrac{kW}{kVA_r}$
    (c) $\cos \theta = \dfrac{kV A}{kW}$
    (d) $\cos \theta = \dfrac{kW}{kVA}$

17. The kVA input of a motor taking 8 kW and 6 kVA$_r$ is
    (a) 8 kVA
    (b) 6 kVA
    (c) 10 kVA
    (d) 14 kVA

18. One reason for high voltage transmission is to
    (a) reduce the length of cable run
    (b) keep the conductor cross-sectional area to a minimum
    (c) allow more cable to be buried underground
    (d) keep the conductor insulation to a minimum

19. The current in the neutral conductor when all three phases are carrying 20 A is
    (a) 0 A
    (b) 20 A
    (c) 40 A
    (d) 60 A

20. The output of the circuit shown in the diagram would be

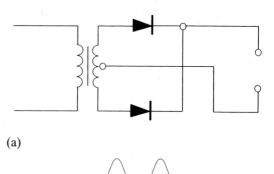

    (a)

    (b)

    (c)

    (d)

Answers

These answers are given for guidance and are not necessarily the only possible solutions.

Chapter 1.

p.1 Bar, horseshoe: By imaginary lines of magnetic flux: Relay, contactor, motor, generator, transformer: An electromotive force (e.m.f.) and induced current

p.3 Try this

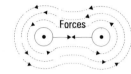

p.4 Try this

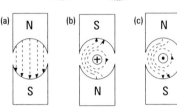

p.4 Try this

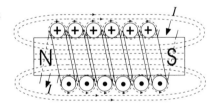

p.5 0.1 T

p.5 1. 1000 At; 2. 0.5 A

p.6 Try this

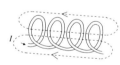

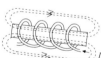

The soft iron core increases the field strength.

p.8 Try this

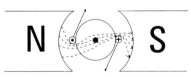

p.8 312.5 A

p.9 6 V

p.9 30 V

p.11 (a) In coil A by self induction, and in coil B by mutual induction due to the flux set up by coil A linking with coil B.

p.12 27 joules

p.13 Try this

p.15 Try this

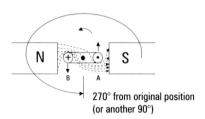

270° from original position
(or another 90°)

p.16 (a)

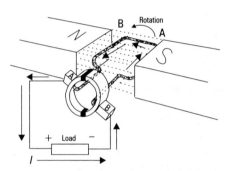

(b) Fleming's R.H. rule, A upwards, inwards, B downwards, outwards. Flows, brush A, load, back, brush B.

p.20 SAQ

1. (a)

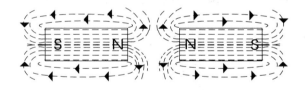

(b) repulsion

2. (a) magnetic flux, (b) magneto-motive force, (c) magnetic flux density

3.6 N

4.333.33 V

5.The primary current sets up a magnetic field, whose lines of flux cut through the secondary winding, hence an e.m.f. is induced into the secondary winding.

6. (a)

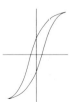

(b) laminate the core

7. When the current flows through the relay's coil it sets up a magnetic field which attracts the armature to the electromagnet's pole-piece, to open and/or close contacts.

8. (a)

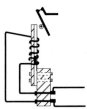

(b) Principle of operation: Excess motor current causes the heater coil temperature to rise, the bimetallic strip heats up and bends towards the trip rod, the trip rod then opens the overload contact at a predetermined point, contactor coil de-energises and contactor drops out.

Chapter 2

p.21 1. 0.133 T; 2. 6.375 V
  3. (a)   (b),

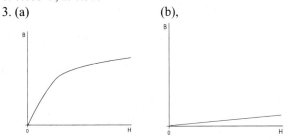

4. P.T.C. type-resistance increases when temperature increases. N.T.C. type-resistance decreases when temperature increases.

p.23 Try this: 1500 C
p.25 Try this: 50 kV/m
p.26 Try this: 3 C/m^2
p.26 Try this: 18.4 mC
p.27 Try this: 21 µF
p.28 Try this: (a) 230 µF; (b) 0.046 C; (c) 0.01 C for 50 µF; 0.016 C for 80 µF, 0.02C for 100 µF
p.29 Try this: (a) 5 µF; (b) 0.5 mC; (c) 0.5 mC
p.29 Try this: 1. 96 µF; 2. 8 µF
p.30 Try this: 0.32 J
p.33 Try this: (a) 10.256 µF; (b) 0.993 J
p.34 SAQ 1. Area of the plates, distance between the plates, type of dielectric. 2. Paper, mica, ceramic, polystyrene, polyester, tantalum oxide (choose any three). 3. 50 µF. 4. (a) volt per metre, (b) coulomb per square metre, (c) coulomb. 5. (a) To improve the power factor of the luminaire. (b) If the maximum working voltage of the capacitor is exceeded it will break down. (c) To safely dissipate the charge and prevent the risk of electric shock. 6. (a) Electric shock long after the mains supply has been isolated, due to stored charge. (b) Use discharge resistors to safely discharge a capacitor's stored charge.

Chapter 3

p.35 1. (a) 2×10^{-6} F; (b) 4×10^{-9} F; (c) 8×10^{-12}. 2. (a) 2 µF; (b) 0.2 mC; (c) 0.2 mC. 3. 0.5 J

(b)

(c) polarised (d) higher values of capacitance for a smaller size
p.36 Try this: 1. 242.5 V. 2. (a) 223.75 V; (b) 41.67 A
p.37 Try this: 1. 344.725 V. 2. (a) 224.9 V; (b) 221.5 V; (c) 213 V
p.40 Try this: 1200 A
p.43 Try this: Maximum current will be taken from the supply. To reduce this current on starting, resistors are connected into the motor circuit.
p.51 SAQ 1. 218.75 V. 2. (a) series wound (b) for speed control
  3. (a)

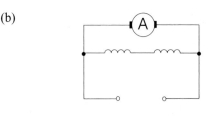

(b)

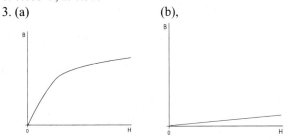

(c)

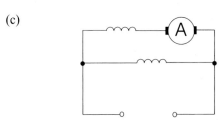

p.52 SAQ 4. To limit the starting current by connecting resistors into the motor circuit. Without this type of starter the starting current would be very excessive.
5. (a) Change either the armature or field connections, not both. (b) change either the armature or both sets of field connections, but not both. 6. 2500A

Chapter 4

p.53 1. 233.5V
  2.

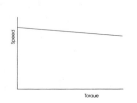

3. (a) overload coil, no-volt release coil. (b) (i) Overload coil operates when excess current flows to return the starting arm to the "off" position. (ii) No-volt coil also returns the starting arm to the "off" position when there is a power failure. 4. (a) Constant speed drives. (b) Fans, conveyors, machine tools (any two)
p.56 Try this: 80 V
p.59 Try this: 1. (a) 29.5 Ω; (b) 7.8 A; (c) (i) 122.5 V, (ii) 195 V

p.59　Try this: 2.

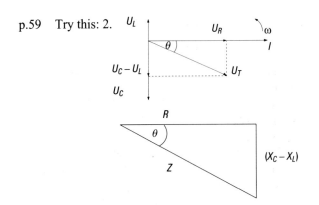

3.

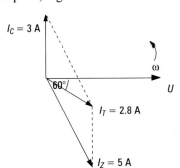

p.60　Try this: 100 Hz
p.61　Try this: 1. (a) 0.5; (b) 60°. 2. (a) 250 V; (b) 53.13°
p.64　Try this: 1578 VA, 0.45 leading, 710 W, 1409 VAr
p.64　Try this: 0.68 lagging
p.67　Try this: (a) 20 A; (b) 276.75 μF
p.68　Try this: 16.3 A
pp. 69 & 70
　　　1. See p. 54, Figure 4.2
　　　2.

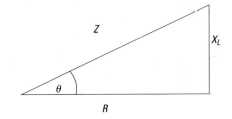

3. (a) 50 Ω; (b) 4.6 A. 4. (a) 0.435 leading; (b) 500 W. 5. (a) lower supply current, smaller cable size, smaller switchgear size, lower cost to the consumer and supply authority (any two); (b) Capacitor connected in parallel with the motor, see p.65, Figure 4.40. 6. 1173 W. 7. (a) 0.887; (b) 27.5°. 8. See p.58, Figure 4.14

Chapter 5
p.71　1.

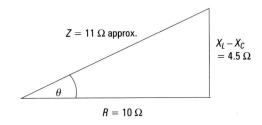

2.

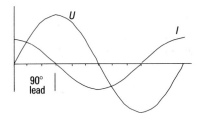

3. (a) 100 V; (b) 207 V
4.

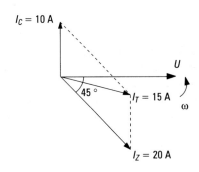

p.75　Try this: (a) 40 kW; (b) 0.8
p.75　Try this: 13.23 kVAr
p.77　Try this: (a) 138.73 V, 245.7 V, 173.4 V; (b) 718 V, 398 V, 190.3 V
p.78　Try this: (a) 57.8 A, 138.7 A, 1156 A; (b) 86.5 A, 415.2 A, 1.73 kA
p.80　Try this: (a) 14.4 kW; (b) 43.2 kW
p.80　Try this: (a) 57.73 A; (b) 33.33 A; (c) 39.995 kW
p.81　Try this: 20 Ω
p.83　Try this: Approx. 13 A
p.88　Try this: (a) Stages 1. To step down the voltage, 2. To convert a.c. to d.c., 3. To smooth the output, 4. To stabilize the output
(b)

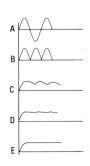

pp. 89 & 90 SAQ 1. Generator 25 kV, Transmission 400/275 kV, Distribution 132/33/11 kV, Commercial 400/230 V, Domestic 230 V
　　　2.

Star				Delta			
U_P	U_L	I_P	I_L	U_P	U_L	I_P	I_L
240 V	415 V	5 A	5 A	415 V	415 V	5 A	8.65 A
110 V	190.5 V	25 A	25 A	110 V	110 V	25 A	43.3 A
219.4 V	380 V	15 A	15 A	380 V	380 V	8.66 A	15 A
138.6 V	240 V	100 A	100 A	240 V	240 V	57.7 A	100 A
680 V	1177.8 V	125 A	125 A	680 V	680 V	125 A	216.5 A

3. Approx. 30.3 A. 4. (a) 82 kVA; (b) 0.68 5. See Figure 5.32. 6. (a) 1.25%; (b) 3%; (c) 4.5%. 7. (a) synchronous speed; (b) supply frequency and the number of pairs of poles

(c)

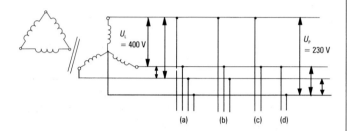

8.

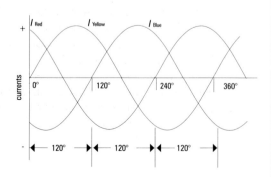

pp 91 & 92: Progress check

1. (a) (i) weber, (ii) tesla; (b) Φ, B. 2. 1 tesla 3. (a) 4 μF; (b) 800 C. 4. (a) area of the plates, distance between the plates, type of dielectric; (b) paper, mica, ceramic, polystyrene, polyester, tantalum oxide (any three). 5. A = series field, B = shunt field, C = armature. 6. To limit the starting current. 7. See p.54, Figure 4.2. 8. (a) (i) 0.68 A, (ii) 0.54 A; (b) power factor correction capacitor. 9. (a) 69.28 A; (b) 40 A; (c) 48 kW. 10. See p.87, Figure 5.32

Chapter 6

p.93   1. (a) I_L = 86.6 A, I_P = 50 A; (b) 60 kW

2.

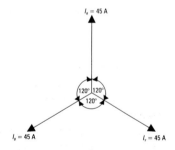

3. See p.87, Figures 5.35 and 5.34 for a and b respectively.

4.

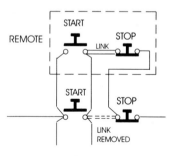

p.96 Try this: 25 rev/s
p.96 Try this: 4%
p.96 (a) 50 rev/s; (b) 48.5 rev/s
p.97 0.04
p.100

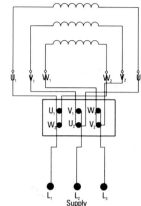

p.112 Try this: See p. 112
p.121 Try this TP & N switches 1, 2 and 3 need "locking off" so that the supply cannot be inadvertently switched back on and cause danger to anyone working on that part of the installation.
pp. 123 & 124 SAQ 1.

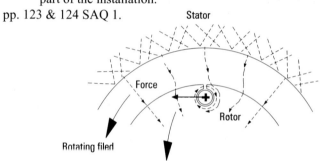

2. (a) 8.33 rev/s; (b) 8 rev/s. 3. (a) refer to p.106, Figure 6.31; (b) to cut out the start winding and capacitor. 4. (a) refer to p.98, Figure 6.8; (b) to reduce "eddy currents". 5. (a) change start or run winding connections NOT BOTH; (b) change any two phase lines; (c) change any two phase lines. 6. (a) (i) wound rotor, (ii) 1. brushes 2. slip rings 3. windings 4. laminated core; (b) to connect the rotor windings to the external resistances

7.

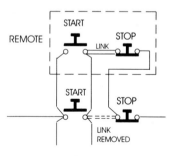

8. (a) embedded in each phase winding of the stator and connected in series; (b) increase in winding temperature – increases thermistor's resistance, this reduces current to contactor coil, therefore it de-energises contactor

Chapter 7

p.125 1. direct-on-line, star-delta, auto-transformer. 2. (a) to produce a movement of flux across the pole pieces; (b) from unshaded to shaded part of the pole face; (c) small fans, display drives, record player turntables, tape decks, office equipment (any two). 3. 6%. 4. (a) to de-energise the contactor when the motor draws excessive current; (b) to prevent automatic restarting of the motor after a supply failure; (c) to prevent motor windings from overheating

p.128 Try this:

$$(a)\ N_p = \frac{I_s}{I_p} \times N_s; \quad (b)\ I_s = \frac{U_p}{U_s} \times I_p$$

$$(c)\ N_s = \frac{U_s}{U_p} \times N_p; \quad (d)\ N_s = \frac{I_p}{I_s} \times N_p$$

p.129 Try this: (a) 4:1; (b) 2000 turns
p.129 Try this: (a) 57.5 V; (b) 0.46
P.133 Try this: 412.5 V
P.134 Try this: 95.65%
p.135 15 kW, 97%
p.139 Try this: 30 VA
pp. 141 & 142 SAQ 1. (a)

(b)

(c)

2. 4.6 kV

3.   (a)

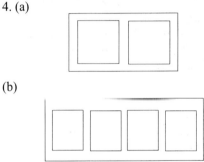

(b)

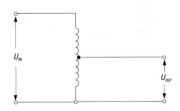

(c)

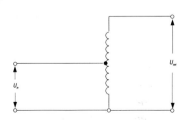

4. 1.3 A. 5. Alternating current in primary – sets up alternating flux in core – which links with turns of secondary winding – inducing a voltage in secondary winding.

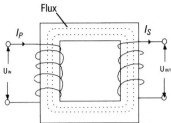

6. (a) to reduce "eddy currents"; (b) short out C.T. secondary before disconnecting ammeter. 7. (a) making and breaking a circuit under ALL conditions; (b) refer to p.140; (c) air, oil, vacuum, air-blast (any three). 8. See p.140, isolator, emergency switch, functional switch purposes

Chapter 8

p. 143 1. 4.6:1 2. 14 kW, 96.6%
3. (a) copper losses change, iron losses remain the same; (b) the area enclosed by the hysteresis loop; (c) when the flux cuts through the core eddy currents are induced
4. (a)

(b)

p.145 Try this: 62.5 lm/W
p.148 Try this: 166.67 lux
p.149 $P_1 = 133.33$ lux, $P_2 = 28.8$ lux
p.150 Try this: 10425.34 lumens
p.163 & 164 SAQ
1. (a) candela; (b) lumens; (c) lumens per watt; (d) lux 2. L.P. sodium, L.P. mercury vapour, coiled coil filament 3. 200 lux 4. $P_1 = 62.5$ lux, $P_2 = 15.26$ lux. 5. Luminaire type, room size, colour and texture of walls and ceilings, number and size of windows, luminaire mounting height (any three). 6. (a) for power factor correction; (b) 1. to start the discharge in the lamp, 2. to limit the current through the lamp, once struck; (c) when closed – heats the tube filaments, when open – open circuits choke to cause H.V. discharge across the lamp. 7. (a) to prolong the life of the filament and maintain the self cleaning cycle; (b) to convert U.V. light to visible light; (c) lamp pressure is too high, lamp must cool so pressure falls before lamp will restrike. 8. See p.161

Chapter 9

p.165 1. (a) luminous intensity; (b) illuminance; (c) luminous flux. 2. comparatively low initial cost, immediate light, good colour rendering, no control gear, easily dimmed (any three). 3. (a) A = choke, B = glow starter, C = tube, D = power factor capacitor; (b) see p.153. 4. 13125 lumens

p.167 Try this: 1. 68 A. 2. (a) 76.9%; (b) 0.43 lagging

p.167 86.2%, 0.85 lagging

p.168 Try this: 3924 N

p.168 Try this: 147.15 kJ

p.169 (a) 1471.5 N; (b) 4414.5 J

p.169 Try this: 18 MJ

p. 170 Try this: 4087.5 W

p.171 Try this: 12262.5 W

pp.172 to 174

SAQ 1. (a) watts; (b) newton metres or joules; (c) newton metres or joules. 2. (a) see p.168; (b) see p.169; (c) see p.166. 3. (a) 2452.5 N; (b) 4905 J. 4. 1471.5 J. 5. 735.75 W. 6. (a) copper, iron, frictional, windage (any two); (b) 6.25 kW. 7. 0.273 kWh. 8. 1600 W

Chapter 10

p. 175 1. Because of the losses that occur within the machine 2. 300 N. 3. 200 kJ. 4. 6.116 kg

p.181 Try this 45 V a.c., 250 V a.c.

p.184 Try this (a) 115W (b) 69W

p.188 SAQ 1. see p. 178. 2. (a) attraction, repulsion; (b) non-linear (uneven). 3. (a) see p.179, Figure 10.9; (b) current measurement without disconnecting circuit or supply

4.

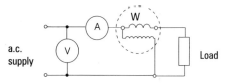

5. (a) energy meter; (b) A = voltage coil, B = current coil. 6. See p.186, Figure 10.30

7.

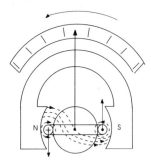

8. (a) no moving parts, clear display, generally more accurate, no parallax error (any two); (b) anti-parallax mirror

pp.189 to 192: End test

1. (a) see p.2, Figure 1.5; (b) repulsion. 2. 4.8 V. 3. 66.67 μF. 4. (a) power factor improvement, smoothing d.c. supplies, starting single-phase induction motors (any two); (b) to discharge the capacitors stored energy safely, and reduce the risk of electric shock. 5. (a) d.c. series wound; (b) speed control; (c) change armature or field connections NOT both. 6. 238.8 V. 7. (a) 100 Ω; (b) 0.6 lagging 8.

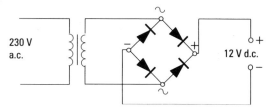

9. (a) 89 kVA; (b) 0.62

10.

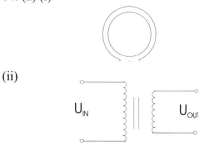

11. (a) to connect the rotor windings to the external resistances; (b) at full speed; (c) change any two phase lines. 12. (a) 16.67 revs/sec; (b) 15.8 revs/sec. 13. (a) 57.5V (b) 0.23

14. (a) (i)

(ii)

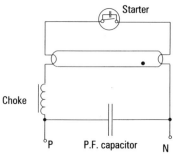

(b) to enable high voltage measurements, on the mains supply, to be taken using a standard voltmeter

15. (a) 12.64 lux; (b) greater

16.

17. (a) 3433.5 N; (b) 10300.5 J. 18. (a) mechanical loss, frictional or windage, electrical loss, copper or iron; (b) 16 kW. 19. (a) (i) to extend the range of a moving coil voltmeter, (ii) to extend the range of a moving coil ammeter; (b) rectifier. 20. See p.186, Figure 10.30

pp. 193 to 194 Multi-choice end test

1. (c); 2. (c); 3. (a); 4. (c); 5. (b); 6. (c); 7. (d); 8. (a); 9. (d); 10. (b); 11. (a); 12. (b); 13. (b); 14. (c); 15. (c); 16. (d); 17. (c); 18. (b); 19. (a); 20. (c)